T0296468

LONDON MATHEMATICAL SOCIETY LECTURE NOTE SERIES

Managing Editor: Professor J.W.S. Cassels, Department of Pure Mathematics and Mathematical Statistics, University of Cambridge, 16 Mill Lane, Cambridge CB2 1SB, England

The titles below are available from booksellers, or, in case of difficulty, from Cambridge University Press.

46 *p*-adic Analysis: a short course on recent work, N. KOBLITZ
59 Applicable differential geometry, M. CRAMPIN & F.A.E. PIRANI
66 Several complex variables and complex manifolds II, M.J. FIELD
86 Topological topics, I.M. JAMES (ed)
87 Surveys in set theory, A.R.D. MATHIAS (ed)
88 FPF ring theory, C. FAITH & S. PAGE
89 An F-space sampler, N.J. KALTON, N.T. PECK & J.W. ROBERTS
90 Polytopes and symmetry, S.A. ROBERTSON
92 Representation of rings over skew fields, A.H. SCHOFIELD
93 Aspects of topology, I.M. JAMES & E.H. KRONHEIMER (eds)
96 Diophantine equations over function fields, R.C. MASON
97 Varieties of constructive mathematics, D.S. BRIDGES & F. RICHMAN
98 Localization in Noetherian rings, A.V. JATEGAONKAR
99 Methods of differential geometry in algebraic topology, M. KAROUBI & C. LERUSTE
100 Stopping time techniques for analysts and probabilists, L. EGGHE
104 Elliptic structures on 3-manifolds, C.B. THOMAS
105 A local spectral theory for closed operators, I. ERDELYI & WANG SHENGWANG
107 Compactification of Siegel moduli schemes, C.-L. CHAI
109 Diophantine analysis, J. LOXTON & A. VAN DER POORTEN (eds)
113 Lectures on the asymptotic theory of ideals, D. REES
114 Lectures on Bochner-Riesz means, K.M. DAVIS & Y.-C. CHANG
116 Representations of algebras, P.J. WEBB (ed)
119 Triangulated categories in the representation theory of finite-dimensional algebras, D. HAPPEL
121 Proceedings of *Groups - St Andrews 1985*, E. ROBERTSON & C. CAMPBELL (eds)
128 Descriptive set theory and the structure of sets of uniqueness, A.S. KECHRIS & A. LOUVEAU
130 Model theory and modules, M. PREST
131 Algebraic, extremal & metric combinatorics, M.-M. DEZA, P. FRANKL & I.G. ROSENBERG (eds)
132 Whitehead groups of finite groups, ROBERT OLIVER
133 Linear algebraic monoids, MOHAN S. PUTCHA
134 Number theory and dynamical systems, M. DODSON & J. VICKERS (eds)
137 Analysis at Urbana, I, E. BERKSON, T. PECK, & J. UHL (eds)
138 Analysis at Urbana, II, E. BERKSON, T. PECK, & J. UHL (eds)
139 Advances in homotopy theory, S. SALAMON, B. STEER & W. SUTHERLAND (eds)
140 Geometric aspects of Banach spaces, E.M. PEINADOR & A. RODES (eds)
141 Surveys in combinatorics 1989, J. SIEMONS (ed)
144 Introduction to uniform spaces, I.M. JAMES
146 Cohen-Macaulay modules over Cohen-Macaulay rings, Y. YOSHINO
148 Helices and vector bundles, A.N. RUDAKOV *et al*
149 Solitons, nonlinear evolution equations and inverse scattering, M. ABLOWITZ & P. CLARKSON
150 Geometry of low-dimensional manifolds 1, S. DONALDSON & C.B. THOMAS (eds)
151 Geometry of low-dimensional manifolds 2, S. DONALDSON & C.B. THOMAS (eds)
152 Oligomorphic permutation groups, P. CAMERON
153 L-functions and arithmetic, J. COATES & M.J. TAYLOR (eds)
155 Classification theories of polarized varieties, TAKAO FUJITA
156 Twistors in mathematics and physics, T.N. BAILEY & R.J. BASTON (eds)
158 Geometry of Banach spaces, P.F.X. MÜLLER & W. SCHACHERMAYER (eds)
159 Groups St Andrews 1989 volume 1, C.M. CAMPBELL & E.F. ROBERTSON (eds)
160 Groups St Andrews 1989 volume 2, C.M. CAMPBELL & E.F. ROBERTSON (eds)
161 Lectures on block theory, BURKHARD KÜLSHAMMER
162 Harmonic analysis and representation theory, A. FIGA-TALAMANCA & C. NEBBIA
163 Topics in varieties of group representations, S.M. VOVSI
164 Quasi-symmetric designs, M.S. SHRIKANDE & S.S. SANE
166 Surveys in combinatorics, 1991, A.D. KEEDWELL (ed)
168 Representations of algebras, H. TACHIKAWA & S. BRENNER (eds)
169 Boolean function complexity, M.S. PATERSON (ed)
170 Manifolds with singularities and the Adams-Novikov spectral sequence, B. BOTVINNIK
171 Squares, A.R. RAJWADE
172 Algebraic varieties, GEORGE R. KEMPF
173 Discrete groups and geometry, W.J. HARVEY & C. MACLACHLAN (eds)
174 Lectures on mechanics, J.E. MARSDEN
175 Adams memorial symposium on algebraic topology 1, N. RAY & G. WALKER (eds)
176 Adams memorial symposium on algebraic topology 2, N. RAY & G. WALKER (eds)
177 Applications of categories in computer science, M. FOURMAN, P. JOHNSTONE & A. PITTS (eds)
178 Lower K- and L-theory, A. RANICKI
179 Complex projective geometry, G. ELLINGSRUD *et al*
180 Lectures on ergodic theory and Pesin theory on compact manifolds, M. POLLICOTT

181 Geometric group theory I, G.A. NIBLO & M.A. ROLLER (eds)
182 Geometric group theory II, G.A. NIBLO & M.A. ROLLER (eds)
183 Shintani zeta functions, A. YUKIE
184 Arithmetical functions, W. SCHWARZ & J. SPILKER
185 Representations of solvable groups, O. MANZ & T.R. WOLF
186 Complexity: knots, colourings and counting, D.J.A. WELSH
187 Surveys in combinatorics, 1993, K. WALKER (ed)
188 Local analysis for the odd order theorem, H. BENDER & G. GLAUBERMAN
189 Locally presentable and accessible categories, J. ADAMEK & J. ROSICKY
190 Polynomial invariants of finite groups, D.J. BENSON
191 Finite geometry and combinatorics, F. DE CLERCK *et al*
192 Symplectic geometry, D. SALAMON (ed)
194 Independent random variables and rearrangement invariant spaces, M. BRAVERMAN
195 Arithmetic of blowup algebras, WOLMER VASCONCELOS
196 Microlocal analysis for differential operators, A. GRIGIS & J. SJÖSTRAND
197 Two-dimensional homotopy and combinatorial group theory, C. HOG-ANGELONI, W. METZLER & A.J. SIERADSKI (eds)
198 The algebraic characterization of geometric 4-manifolds, J.A. HILLMAN
199 Invariant potential theory in the unit ball of $\mathbf{C}^n$, MANFRED STOLL
200 The Grothendieck theory of dessins d'enfant, L. SCHNEPS (ed)
201 Singularities, JEAN-PAUL BRASSELET (ed)
202 The technique of pseudodifferential operators, H.O. CORDES
203 Hochschild cohomology of von Neumann algebras, A. SINCLAIR & R. SMITH
204 Combinatorial and geometric group theory, A.J. DUNCAN, N.D. GILBERT & J. HOWIE (eds)
205 Ergodic theory and its connections with harmonic analysis, K. PETERSEN & I. SALAMA (eds)
206 An introduction to noncommutative differential geometry and its physical applications, J. MADORE
207 Groups of Lie type and their geometries, W.M. KANTOR & L. DI MARTINO (eds)
208 Vector bundles in algebraic geometry, N.J. HITCHIN, P. NEWSTEAD & W.M. OXBURY (eds)
209 Arithmetic of diagonal hypersurfaces over finite fields, F.Q. GOUVÊA & N. YUI
210 Hilbert C*-modules, E.C. LANCE
211 Groups 93 Galway / St Andrews I, C.M. CAMPBELL *et al* (eds)
212 Groups 93 Galway / St Andrews II, C.M. CAMPBELL *et al* (eds)
214 Generalised Euler-Jacobi inversion formula and asymptotics beyond all orders, V. KOWALENKO, N.E. FRANKEL, M.L. GLASSER & T. TAUCHER
215 Number theory 1992–93, S. DAVID (ed)
216 Stochastic partial differential equations, A. ETHERIDGE (ed)
217 Quadratic forms with applications to algebraic geometry and topology, A. PFISTER
218 Surveys in combinatorics, 1995, PETER ROWLINSON (ed)
220 Algebraic set theory, A. JOYAL & I. MOERDIJK
221 Harmonic approximation, S.J. GARDINER
222 Advances in linear logic, J.-Y. GIRARD, Y. LAFONT & L. REGNIER (eds)
223 Analytic semigroups and semilinear initial boundary value problems, KAZUAKI TAIRA
224 Computability, enumerability, unsolvability, S.B. COOPER, T.A. SLAMAN & S.S. WAINER (eds)
225 A mathematical introduction to string theory, S. ALBEVERIO, J. JOST, S. PAYCHA, S. SCARLATTI
226 Novikov conjectures, index theorems and rigidity I, S. FERRY, A. RANICKI & J. ROSENBERG (eds)
227 Novikov conjectures, index theorems and rigidity II, S. FERRY, A. RANICKI & J. ROSENBERG (eds)
228 Ergodic theory of $\mathbf{Z}^d$ actions, M. POLLICOTT & K. SCHMIDT (eds)
229 Ergodicity for infinite dimensional systems, G. DA PRATO & J. ZABCZYK
230 Prolegomena to a middlebrow arithmetic of curves of genus 2, J.W.S. CASSELS & E.V. FLYNN
231 Semigroup theory and its applications, K.H. HOFMANN & M.W. MISLOVE (eds)
232 The descriptive set theory of Polish group actions, H. BECKER & A.S. KECHRIS
233 Finite fields and applications, S. COHEN & H. NIEDERREITER (eds)
234 Introduction to subfactors, V. JONES & V.S. SUNDER
235 Number theory 1993–94, S. DAVID (ed)
236 The James forest, H. FETTER & B. GAMBOA DE BUEN
237 Sieve methods, exponential sums, and their applications in number theory, G.R.H. GREAVES, G. HARMAN & M.N. HUXLEY (eds)
238 Representation theory and algebraic geometry, A. MARTSINKOVSKY & G. TODOROV (eds)
239 Clifford algebras and spinors, P. LOUNESTO
240 Stable groups, FRANK O. WAGNER
241 Surveys in combinatorics, 1997, R.A. BAILEY (ed)
242 Geometric Galois actions I, L. SCHNEPS & P. LOCHAK (eds)
243 Geometric Galois actions II, L. SCHNEPS & P. LOCHAK (eds)
244 Model theory of groups and automorphism groups, D. EVANS (ed)
245 Geometry, combinatorial designs and related structures, J.W.P. HIRSCHFELD, S.S. MAGLIVERAS & M.J. DE RESMINI (eds)
246 *p*-Automorphisms of finite *p*-groups, E.I. KHUKHRO
247 Analytic number theory, Y. MOTOHASHI (ed)
248 Tame topology and o-minimal structures, LOU VAN DEN DRIES
249 The atlas of finite groups: ten years on, ROBERT CURTIS & ROBERT WILSON (eds)
250 Characters and blocks of finite groups, G. NAVARRO
251 Gröbner bases and applications, B. BUCHBERGER & F. WINKLER (eds)
252 Geometry and cohomology in group theory, P. KROPHOLLER, G. NIBLO, R. STÖHR (eds)

London Mathematical Society Lecture Note Series. 248

Tame Topology and O-minimal Structures

Lou van den Dries
University of Illinois at Urbana-Champaign

PUBLISHED BY THE PRESS SYNDICATE OF THE UNIVERSITY OF CAMBRIDGE
The Pitt Building, Trumpington Street, Cambridge CB2 1RP, United Kingdom

CAMBRIDGE UNIVERSITY PRESS
The Edinburgh Building, Cambridge, CB2 2RU, United Kingdom
40 West 20th Street, New York, NY 10011-4211, USA
10 Stamford Road, Oakleigh, Melbourne 3166, Australia

First published 1998

A catalogue record for this book is available from the British Library

ISBN 0 521 59838 9 paperback

Transferred to digital printing 2003

CONTENTS

PREFACE

My aim is to show that o-minimal structures provide an excellent framework for developing tame topology, or *topologie modérée*, as outlined in Grothendieck's prophetic "Esquisse d'un Programme" of 1984. This close connection between tame topology and a subject created by model-theorists is hardly controversial, though perhaps not yet widely known or understood.

In the early 1980s I had noticed that many properties of semialgebraic sets and maps could be derived from a few simple axioms, essentially the axioms defining "o-minimal structures", as their models came to be called in an influential article by Pillay and Steinhorn. After Wilkie established in 1991 that the exponential field of real numbers is o-minimal the subject has grown rapidly. The supply of o-minimal structures on the real field is still increasing. In combination with general o-minimal finiteness theorems this gives rise to applications in real algebraic and real analytic geometry.

A rough version of this book was circulated informally in 1991, and was based on articles by various authors and on courses I gave at Stanford, Konstanz, and the University of Illinois at Urbana-Champaign. The main additions since then consist of Chapter 5 about the Vapnik-Chervonenkis property, Section 2 of Chapter 6 on fiberwise properties, Section 3 of Chapter 9 on a conjecture of Benedetti and Risler, and Chapter 10 on definable spaces. I have taken pains to present the subject in a way that is widely accessible and requires no knowledge of model theory. My initial ambition was to include more on the construction of o-minimal structures, but I decided to postpone this. (Traces remain as occasional references to what is supposed to happen "in the next volume".) Some material is included explicitly in the form of digressions or exercises requiring further background on the reader's part. Also, in solving a problem of Benedetti and Risler I quote without proof Wilkie's theorem of 1991. Otherwise everything is developed from scratch. Each chapter has notes at the end with references and comments.

It is a pleasure to thank many people for their careful reading of parts of earlier versions, their suggestions and their interest: M. Aschenbrenner, W. Henson, J. Holly, J. Iovino, G. Kreisel, A. Lewenberg, A. Macintyre, D. Marker, C. Miller, A. Nesin, S. Schanuel, P. Speissegger, S. Świerczkowski, A. Wilkie, and A. Woerheide.

The author is grateful for support by the National Science Foundation during the years this book was in the making. I also thank J. Finkler for converting the text into TEX.

May 1997 Lou van den Dries, Urbana, Illinois

Prerequisites

I tried to keep these minimal without compromising a natural development of the subject. This book only requires familiarity with

From algebra. Basic properties of groups, rings (always associative with 1), fields, polynomials, linear maps and their matrices and determinants.

From topology. The notions of (Hausdorff) topological space, closure and interior, continuity of functions between topological spaces, connectedness.

From analysis. Elementary calculus of real-valued functions of real arguments. In Chapter 2 we use at one point the argument principle from complex analysis.

The logical dependence of chapters on earlier ones is roughly as follows:

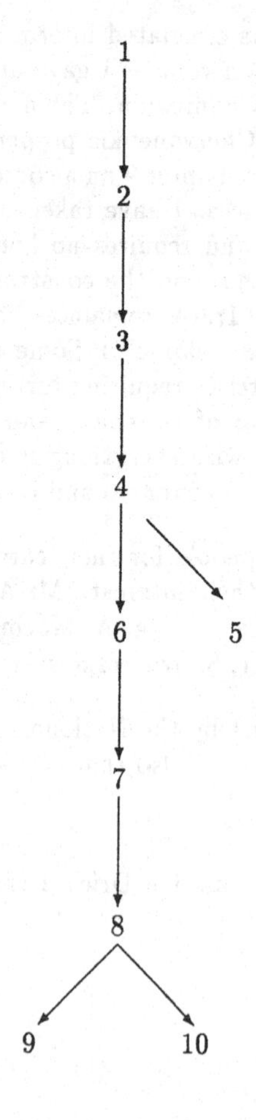

Conventions and Notations Used Throughout This Book

Our notations are fairly standard, but it may be useful to list some of them.

$\mathbb{N}$, $\mathbb{Z}$, $\mathbb{Q}$, $\mathbb{R}$, and $\mathbb{C}$ denote the sets of natural numbers (= nonnegative integers), integers, rational numbers, real numbers, and complex numbers, respectively. We always let k, m, n (possibly with accents or subscripts) range over $\mathbb{N} = \{0, 1, 2, \ldots\}$.

$\ldots := \ldots\ldots$	$\ldots$ is by definition equal to $\ldots\ldots$;
$A \subseteq B$	A is a subset of the set B;
$A - B$	the relative complement $\{x \in A : x \notin B\}$ of the set B in the set A;
$f : A \to B$	f is a map from the set A into the set B.

Formally, a map $f : A \to B$ is an ordered triple, consisting of

(i) $\Gamma(f) := \{(a, f(a)) : a \in A\}$ (the **graph** of f, a subset of $A \times B$),
(ii) the set A (the **domain** of f),
(iii) the set B (the **codomain** of f).

Given a map $f : A \to B$ and a set $B' \subseteq B$ with $f(A) \subseteq B'$ we indicate the map $a \mapsto f(a) : A \to B'$ by $f : A \to B'$, and similarly, if $B \subseteq B''$ we indicate the map $a \mapsto f(a) : A \to B''$ by $f : A \to B''$. (This may be an abuse of notation but is convenient and will not lead to confusion.) Also, with f as above and $A' \subseteq A$ we let $f|A' : A' \to B$ be the restriction of f to A'. "Function" is used as a synonym for "map". Usually we write "function" if we think of its values as numerical.

"Collection" is a synonym for "set"; we use "collection" if we think of its members as sets themselves. We do maintain the distinction between a **family** $(a_i)_{i \in I}$, which is formally the assignment $i \mapsto a_i$, and the set $\{a_i : i \in I\}$ of values of this assignment.

We let $\mathcal{P}(X)$ denote the power set of the set X, and let $|X|$ be the cardinality of the (usually finite) set X. (But we also write $|x|$ for the norm of a vector x, and $|K|$ for the polyhedron spanned by a complex K; precise definitions of $|x|$ and $|K|$ are given in the relevant chapters. We trust that this multiple use of the notation $|\cdot|$ will not lead to confusion.)

Given a set R we let R^n be the set of all n-tuples $(x_1, \ldots, x_n)$ with all $x_i \in R$. Thus R^0 contains exactly one element, the empty tuple 0, so $R^0 = \{0\}$.

If $S \subseteq A \times B$ and $a \in A$ we put $S_a := \{b \in B : (a, b) \in S\}$, and we view S as describing the family $(S_a)_{a \in A}$ of subsets of B. Usually this situation occurs where a set R is given, and $A = R^m$, $B = R^n$, and $A \times B$ is identified with R^{m+n}.

A **partition** of a set S is a collection of *nonempty* pairwise disjoint subsets of S whose union is S. A partition $\mathcal{P}$ of S is said to be **compatible** with a subset $X \subseteq S$ (or is said to **partition** X) if X is a union of sets in $\mathcal{P}$; in particular, $\mathcal{P}$ is compatible with the empty subset of S. A **refinement** of a partition $\mathcal{P}$ of S is a partition $\mathcal{P}'$ of S that is compatible with each set in $\mathcal{P}$.

Let T be a topological space and $S \subseteq T$. If the ambient space T is clear from the context then $\text{cl}(S)$ indicates the **closure** of S in T, and $\text{int}(S)$ the **interior** of S in T. Also $\text{bd}(S) := \text{cl}(S) - \text{int}(S)$ denotes the **boundary** of S in T, and $\partial S := \text{cl}(S) - S$ denotes the **frontier** of S in T. (Some books use other terminology or notation.) If T is not clear from the context we write $\text{cl}_T(S)$ and $\text{int}_T(S)$. We say that X is an open (respectively, closed) subset of S if X is a subset of S and X is open (respectively, closed) in the induced topology on S. Given a map $f: S \to H$ from S into a Hausdorff space H and given a point $p \in \text{cl}(S \setminus \{p\})$ there is clearly at most one point $q \in H$ such that for each neighborhood V of q there is a neighborhood U of p such that $f(U \cap (S \setminus \{p\})) \subseteq V$. We write

$$\lim_{x \to p} f(x) = q$$

to indicate that q is such a point. (If in addition $p \in S$ and f is continuous at p, then this just means $f(p) = q$.)

INTRODUCTION AND OVERVIEW

Consider "nice" subsets of Euclidean spaces, defined by conditions like

$$(*) \qquad f_1(x) = f_2(x) = \cdots = f_k(x) = 0, \quad g_1(x) > 0, \ldots, g_l(x) > 0,$$

$x = (x_1, \ldots, x_n) \in \mathbb{R}^n$, for "nice" functions $f_1, \ldots, f_k, g_1, \ldots, g_l$. Now perform elementary logical, geometric and topological operations on these sets: take unions, intersections, complements, closures and cartesian products, and project into lower-dimensional euclidean spaces. If we keep repeating these operations with the new sets that arise, then, roughly speaking, two kinds of things can happen:

(1) After only a few such operations, say after taking projections of finite unions of the sets one starts out with, a *stabilization* occurs, and performing further operations does not produce any new sets.

(2) Ever more complicated sets arise: Cantor-like sets, Borel sets of arbitrarily high complexity, and so on. (For an example, see Chapter 1, (2.6).)

It is phenomenon (1) that is of particular interest to us: here we remain in the realm of geometry and topology envisaged by Poincaré, and we may view the study of the corresponding category of "nice" spaces and maps between them as an unfolding of the rich algebraic-analytic-topological structure of the continuum.

When phenomenon (2), on the other hand, is pursued, then the ties to geometry tend to become very loose. We enter here *Cantor's paradise*: only set-theoretic features ultimately survive in this bleak landscape. Already on the Borel level the real line and the real plane (or Cantor space for that matter) become indistinguishable.

Despite these large contrasts it is often hard to determine for a specific class of "nice functions" whether we are in case (1) or in case (2). In Chapter 1 we show that with linear functions we are in case (1), and in Chapter 2 we do the same for polynomial functions. But only in a later volume shall we pay special attention to the art of proving that many interesting situations fall under phenomenon (1). **The emphasis in this book is on developing the general theory of what else is true if one happens to be in case (1).**

A key example of phenomenon (1) is provided by the semialgebraic sets. A **semialgebraic** set in $\mathbb{R}^n$ is a finite union of sets of the form $(*)$ where $f_1, \ldots, f_k, g_1, \ldots, g_l \in \mathbb{R}[X]$, $X = (X_1, \ldots, X_n)$. (If the polynomials f_i and g_j are also assumed to be of degree ≤ 1 we speak of **semilinear** sets.) In particular, for $n = 1$, we get just the finite unions of intervals and points, or equivalently, the subsets of the real line with only finitely many connected components.

A major fact about semialgebraic sets is the Tarski-Seidenberg projection property:

If $A \subseteq \mathbb{R}^{n+1}$ is semialgebraic, then $\pi(A) \subseteq \mathbb{R}^n$ is semialgebraic, where $\pi: \mathbb{R}^{n+1} \to \mathbb{R}^n$ is the projection map on the first n coordinates.

Related results are: the closure, interior, and convex hull of a semialgebraic set are semialgebraic; a semialgebraic set has only finitely many connected components, each of which is also semialgebraic; a semialgebraic set can be triangulated, and stratified into finitely many semialgebraic real analytic manifolds. (All of this remains true with "semilinear" instead of "semialgebraic".) A proper setting for many results of this kind is provided by the theory of o-minimal structures, as this book will attempt to show.

O-minimal structures on the real line

The class of semialgebraic sets and the class of semilinear sets, are both examples of an o-minimal structure on $\mathbb{R}$, the ordered set of real numbers (the real line).

DEFINITION. An **o-minimal structure on $\mathbb{R}$** is a sequence $\mathcal{S} = (\mathcal{S}_n)_{n \in \mathbb{N}}$ such that for each n:

(1) $\mathcal{S}_n$ is a boolean algebra of subsets of $\mathbb{R}^n$, that is, $\mathcal{S}_n$ is a collection of subsets of $\mathbb{R}^n$, $\emptyset \in \mathcal{S}_n$ and if $A, B \in \mathcal{S}_n$, then $A \cup B \in \mathcal{S}_n$, and $\mathbb{R}^n - A \in \mathcal{S}_n$;
(2) $A \in \mathcal{S}_n \Rightarrow A \times \mathbb{R} \in \mathcal{S}_{n+1}$ and $\mathbb{R} \times A \in \mathcal{S}_{n+1}$;
(3) $\{(x_1, \ldots, x_n) \in \mathbb{R}^n : x_i = x_j\} \in \mathcal{S}_n$ for $1 \leq i < j \leq n$;
(4) $A \in S_{n+1} \Rightarrow \pi(A) \in \mathcal{S}_n$, where $\pi: \mathbb{R}^{n+1} \to \mathbb{R}^n$ is the usual projection map;
(5) $\{r\} \in \mathcal{S}_1$ for each $r \in \mathbb{R}$, and $\{(x, y) \in \mathbb{R}^2 : x < y\} \in \mathcal{S}_2$;
(6) the only sets in $\mathcal{S}_1$ are the finite unions of intervals and points. ("Interval" always means "open interval", with infinite endpoints allowed.)

The first four axioms guarantee that each set $A \subseteq \mathbb{R}^n$ that is definable (in a precise logical sense) from the sets in $\mathcal{S}$ also belongs to $\mathcal{S}$. (See Chapter 1.) With axiom (5) they imply that finite unions of intervals and points belong to $\mathcal{S}_1$. Axiom (6) says that no further subsets of $\mathbb{R}$ belong to $\mathcal{S}_1$; this is the minimality axiom that explains the term "o-minimal", "o" standing for "order". This o-minimality axiom can be viewed as expressing compatibility with the order (and hence topology) of the real line, and is clearly the simplest compatibility condition of this nature.

Examples of o-minimal structures on the real line

- the semilinear sets;
- the semialgebraic sets;
- the subsets of the affine spaces $\mathbb{R}^n$ for $n = 0, 1, 2, \ldots$ that are subanalytic in the larger projective space $\mathbb{P}^n(\mathbb{R})$;

- the images in $\mathbb{R}^n$ for $n = 0, 1, 2, \ldots$ under projection maps $\mathbb{R}^{n+k} \to \mathbb{R}^n$ of sets of the form $\{(x,y) \in \mathbb{R}^{n+k} : P(x,y,e^x,e^y) = 0\}$ where P is a real polynomial in $2(n+k)$ variables, and where $x = (x_1, \ldots, x_n)$, $y = (y_1, \ldots, y_k)$, $e^x = (e^{x_1}, \ldots, e^{x_n})$, $e^y = (e^{y_1}, \ldots, e^{y_k})$.

These examples and others will be treated in some detail in this book and its planned sequel: the first two in Chapters 1 and 2 of this book, the others in a later volume. The o-minimality of the last example follows from a remarkable theorem proved by Wilkie in 1991. In contrast to the first three examples it contains the graph of a function on $\mathbb{R}$ growing faster at ∞ than any polynomial, namely e^x. There are many other interesting o-minimal structures on $\mathbb{R}$, but not too many. For instance, a result of Peterzil [45] says that there is exactly one o-minimal structure on $\mathbb{R}$ that lies strictly between the class of semilinear sets and the class of semialgebraic sets; this intermediate structure contains every bounded semialgebraic set, but not the graph of multiplication.

General facts about arbitrary o-minimal structures on the real line

From now on in this introduction we fix an o-minimal structure $\mathcal{S}$ on $\mathbb{R}$.

Let $A \subseteq \mathbb{R}^m$ and $f: A \to \mathbb{R}^n$. We say A is **definable** if $A \in \mathcal{S}_m$. The map f is said to be **definable** if its graph $\Gamma(f) \subseteq \mathbb{R}^{m+n}$ is definable. If f is definable, then the domain A of f and its image $f(A) \subseteq \mathbb{R}^n$ are also definable. The closure $\text{cl}(A)$ and interior $\text{int}(A)$ of a definable set $A \subseteq \mathbb{R}^n$ are definable. There are many more such elementary but useful facts, and Chapter 1 is basically about them. A less easy but fundamental result, proved in Chapter 3, is the following:

MONOTONICITY THEOREM.

If the function $f: (a,b) \to \mathbb{R}$ is definable, then there are $a = a_0 < a_1 < \cdots < a_N = b$ such that f is continuous on each subinterval (a_i, a_{i+1}), and either constant, or strictly increasing, or strictly decreasing, on each subinterval (a_i, a_{i+1}).

This result is used throughout the subject. In particular it plays a key role in the proof that definable sets can be partitioned into finitely many *cells*. Cells are nonempty definable sets of an especially simple form. They are defined inductively as follows:

(i) the cells in $\mathbb{R}^1 = \mathbb{R}$ are just the points $\{r\}$, and the intervals (a,b);
(ii) let $C \subseteq \mathbb{R}^n$ be a cell; if $f, g: C \to \mathbb{R}$ are definable continuous functions such that $f < g$ on C, then

$$(f,g) := \{(x,r) \in C \times \mathbb{R} : f(x) < r < g(x)\}$$

is a cell in $\mathbb{R}^{n+1}$; also, given a definable continuous function $f: C \to \mathbb{R}$ the sets

$$\Gamma(f),$$
$$(-\infty, f) := \{(x,r)\in C\times\mathbb{R}:\ r < f(x)\},$$
$$(f, +\infty) := \{(x,r)\in C\times\mathbb{R}:\ f(x) < r\},$$

are cells in $\mathbb{R}^{n+1}$; finally $C\times\mathbb{R}\subseteq\mathbb{R}^{n+1}$ is a cell. (See figure.)

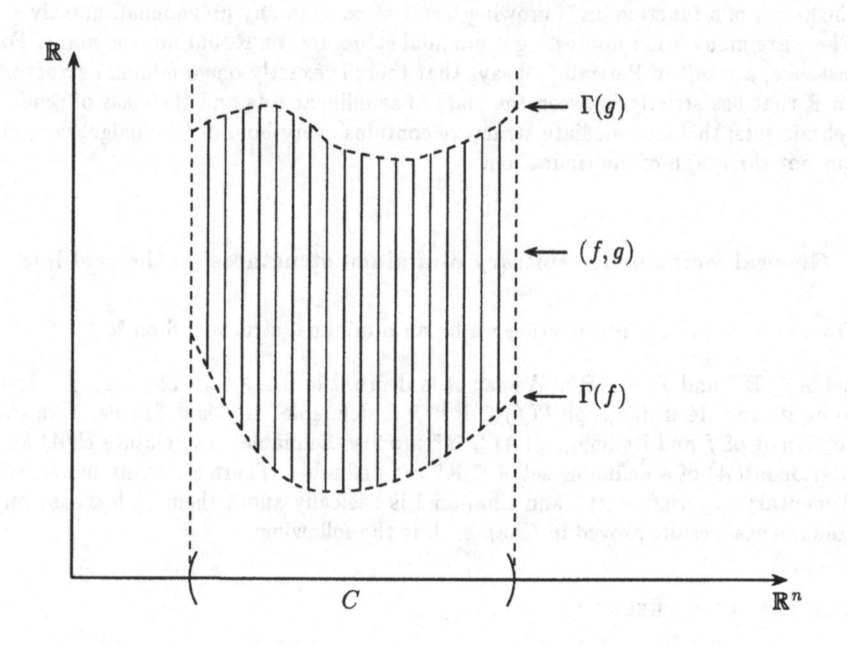

It is extremely useful in inductive proofs to view the cells defined in (ii) as fiber spaces over their projection C, and in general to view $\mathbb{R}^{n+1}$ as fibered over $\mathbb{R}^n$ via the projection map $\pi:\mathbb{R}^{n+1}\to\mathbb{R}^n$ on the first n coordinates. The next result is also established in Chapter 3 and is central in the subject. It is to some extent the many-variable version of the Monotonicity Theorem.

CELL DECOMPOSITION THEOREM.

Each definable set $A\subseteq\mathbb{R}^m$ has a finite partition $A = C_1\cup\cdots\cup C_k$ into cells C_i. If $f: A\to\mathbb{R}^n$ is a definable map, this partition of A can moreover be chosen such that all restrictions $f|C_i$ are continuous.

It follows in particular that a definable set has only finitely many connected components, and these components are again definable (and path connected).

Definable invariants (Chapter 4)

The **dimension** of a cell is defined inductively in an obvious way, and then extended to all definable sets $A \subseteq \mathbb{R}^m$ by $\dim A := \max\{\dim(C) : C \subseteq A \text{ is a cell}\}$ if A is nonempty, while $\dim(\emptyset) = -\infty$. This dimension has good properties:

(1) *Let $f: A \to \mathbb{R}^n$ be a definable map, not necessarily continuous. Then* $\dim A \geq \dim f(A)$; *in particular* $\dim A = \dim f(A)$ *if f is injective.*
(2) $\dim(\text{cl}(A) - A) < \dim A$ for nonempty definable $A \subseteq \mathbb{R}^m$.

A more subtle "definable invariant" is the **Euler characteristic**. For a cell C of dimension d it takes the value $E(C) := (-1)^d$, and for an arbitrary definable set A we put

$$E(A) := \sum E(C_i)$$

where $C_1, \ldots, C_k$ are the members of a finite partition of A into cells. One proves easily that this definition makes sense. More difficult is the following:

If $f: A \to \mathbb{R}^n$ is an injective definable map (not necessarily continuous!), then

$$E(A) = E\big(f(A)\big).$$

Results on definable families and definable collections (Chapters 4, 5)

Let $S \subseteq \mathbb{R}^{m+n}$ be a definable set. Then S gives rise to a family $(S_x)_{x \in \mathbb{R}^m}$ of definable subsets of $\mathbb{R}^n$, where $S_x := \{y \in \mathbb{R}^n : (x, y) \in S\}$. We call this a **definable family** (with parameter space $\mathbb{R}^m$), and we are interested in how the properties of S_x depend on x. Cell decomposition gives a uniform bound on the number of cells into which S_x can be partitioned, independent of x, and a more precise analysis along these lines leads to:

(1) *For each $d \in \{0, \ldots, n\}$ the set $S(d) := \{x \in \mathbb{R}^m : \dim(S_x) = d\}$ is definable and*

$$\dim\big\{(x, y) \in S : \dim(S_x) = d\big\} = \dim\big(S(d)\big) + d;$$

(2) *$E(S_x)$ takes only finitely many values as x runs through $\mathbb{R}^m$. Moreover, for each integer e the set $A(e) := \big\{x \in \mathbb{R}^m : E(S_x) = e\big\}$ is definable, and*

$$E\big\{(x, y) \in S : x \in A(e)\big\} = e \cdot E\big(A(e)\big).$$

The collection of sets $\mathcal{C} := \{S_x : x \in \mathbb{R}^m\}$ is called a **definable collection.** (We distinguish it from the **family** $(S_x)_{x \in \mathbb{R}^m}$.) This collection has a surprising combinatorial property, as we explain now.

For each finite set $F \subseteq \mathbb{R}^n$, put $F \cap C := \{F \cap S_x : x \in \mathbb{R}^m\}$, so $0 \leq |F \cap C| \leq 2^{|F|}$. Now define a growth function $g_S : \mathbf{N} \to \mathbf{N}$ by

$$g_S(k) := \max\{|F \cap C| : F \text{ is a } k\text{-element subset of } \mathbb{R}^n\},$$

so that $0 \leq g_S(k) \leq 2^k$. A remarkable fact is:

(3) *The function $g_S(k)$ is bounded by a polynomial in k.*

(In probabilistic terms this means that $\{S_x : x \in \mathbb{R}^m\}$ is a VC-class or Vapnik-Chervonenkis class; when $\mathbb{R}^n$ is equipped with a probability measure it implies that the law of large numbers holds uniformly for all sets S_x, with bounds that are independent of x.)

All the above actually goes through when instead of the real line $\mathbb{R}$ we take any dense linearly ordered set without endpoints. The notion of o-minimal structure on such an ordered set is defined in the same way as for $\mathbb{R}$. (A basis for the topology on the ordered set is given by the intervals (a, b), with the product topology on cartesian powers; the usual notion of "connected" is replaced by "definably connected", which makes no difference when the underlying ordered set is $\mathbb{R}$.) This kind of generalization may seem at first unmotivated, but is actually crucial to prove certain results in the classical real case. We discuss this later in the introduction.

The case that $\mathcal{S}$ contains addition and multiplication (Chapters 6–10)

Up till now we have considered a completely arbitrary o-minimal structure $\mathcal{S}$ on $\mathbb{R}$, and we did not even assume that $\mathcal{S}$ contains (the graph of) addition. *Assume $\mathcal{S}$ contains addition.* Then we have a very useful definable choice principle based on the possibility of singling out the midpoint of a bounded interval. Special cases of this general principle are the "curve selection lemma", and the fact that a definable equivalence relation on a definable set has a definable set of representatives. These facts have numerous consequences (see Chapter 6) but for deeper results we need also the presence of multiplication. *Assume $\mathcal{S}$ also contains multiplication.*

Then in the cell decomposition theorem we can require the cells to be C^1-manifolds, and even C^2-manifolds, and so on (see Chapter 7). (However, for all I know it may be that finer and finer partitions are required as the desired degree of smoothness is increased.) We use C^1-cell decomposition to prove a "good directions lemma":

Given a definable set $A \subseteq \mathbb{R}^{n+1}$ of dimension $< n+1$ there is a unit vector $u \in S^n$ such that every affine line in $\mathbb{R}^{n+1}$ with direction u intersects A in only finitely many points.

This is a substitute for Noether normalization in the proof of the following fundamental result in Chapter 8.

TRIANGULATION THEOREM. *Each definable set is definably homeomorphic to a bounded semilinear set.*

One curious consequence of the triangulation theorem is that two definable sets are definably equivalent (that is, there is a definable, not necessarily continuous, bijection between them) if and only if they have the same dimension and the same Euler characteristic. There are numerous other consequences, such as a definable version of the Tietze extension lemma (Chapter 8, Section 3).

Call a definable map $f: E \to B$ between definable sets **definably trivial** if there are a definable set F and a definable homeomorphism $E \to B \times F$ such that the diagram

$$\begin{array}{ccc} E & \xrightarrow{\sim} & B \times F \\ & {\scriptstyle f}\searrow \quad \swarrow & \\ & B & \end{array}$$

commutes. (Note that then all fibers $f^{-1}(b)$ are definably homeomorphic to F.) In Chapter 9 we prove

TRIVIALIZATION THEOREM. *Let $f: E \to B$ be a definable continuous map between definable sets E and B. Then B can be partitioned into definable sets $B_1, \ldots, B_k$ such that the restrictions*

$$f|f^{-1}(B_i): f^{-1}(B_i) \to B_i, \ i = 1, \ldots, k,$$

are definably trivial.

As a corollary, a definable collection $\{S_x : x \in \mathbb{R}^m\}$ of definable sets contains only finitely many definable homeomorphism types. Another standard consequence is the conical structure of a definable set in some ball around any of its points.

In combination with the o-minimality of the exponential field of real numbers (Wilkie's theorem) a variant of the trivialization theorem leads to the proof of a conjecture by Benedetti and Risler (Chapter 9, Section 3).

In the final chapter, Chapter 10, we show that under rather weak assumptions definable sets can be glued together by means of definable continuous transition maps. We also use the triangulation theorem to form quotient spaces in the category of definable sets and definable continuous maps.

The triangulation and trivialization theorems go through for each real closed field equipped with an o-minimal structure that contains the addition and multiplication of the field. In fact we prove these results directly in this setting, adapting the treatment of the semialgebraic case by Delfs and Knebusch [11]. This kind of generalization is done not for its own sake but to throw light on the classical real case: certain real closed extensions of $\mathbb{R}$, like the field of algebraic Puiseux series over $\mathbb{R}$, appear naturally as real closures of function fields of real algebraic varieties. More generally, if $\mathbb{R}$ and a real closed extension R are equipped with corresponding o-minimal structures, then points in R^n serve as generic points of definable subsets of $\mathbb{R}^n$. (This idea will only play a minor role here, but will become quite useful in a later volume where the model-theoretic tools to handle such situations are made available.) One could also argue that some real closed fields, like the field of real algebraic numbers and, at the other extreme, Conway's field of surreal numbers, are of independent interest. (In contrast to real closed fields in general, Conway's field perhaps supports a *canonical* exponential function for which Wilkie's theorem on exponentiation goes through.)

This finishes our preview of the main results. We actually prove stronger and more detailed versions of these theorems, and we also include exercises developing some material a little further. To read this book only a rudimentary knowledge of real analysis and point set topology is needed. In particular, no model theory is used, although the subject of o-minimal structures developed in close contact with model theory. The deeper results that will be treated in a later volume do require some elementary results from model theory, valuation theory, and differential topology.

Notes, comments, and references

Artin and Schreier [1], Koopman and Brown [36] and Tarski [61] are at the origin of much of our subject. Standard references for real algebraic and semialgebraic geometry are the books by Brumfiel [7], Bochnak, Coste and Roy [4], and Benedetti and Risler [2]; see also the articles by Bochnak and Efroymson [5], Delfs and Knebusch [11], and Knebusch [34]. The theory of semianalytic and subanalytic sets was created by Łojasiewicz [40], Gabrielov [26] and Hironaka [30]; see also Bierstone and Milman [3], Denef and Van den Dries [15], and Sussmann [60], for expositions and alternative treatments. Grothendieck's proposed *topologie modérée* in [28] would contain all those theories as special cases and seems very much in the spirit of o-minimality. The appendix in MacPherson [42] also puts forward a general view on tame topology. It was observed in [20] that the theory of subanalytic sets falls under the scope of o-minimality; see also Van den Dries and Miller [23] for more on this, and for an exposition of o-minimality in the more geometric language of manifolds.

Arbitrary o-minimal structures on the real line were first considered in [19], where they were called "structures of finite type". The term "o-minimal structure" was suggested by Pillay and Steinhorn [49], because of an analogy with the notion of "strongly minimal structure" from model theory. This analogy has since turned out to be very suggestive though it plays no role in this book. The article [49] and its sequels [35] and [48] systematically develop the theory of o-minimal structures on arbitrary linearly ordered sets from a model-theoretic viewpoint. This direction found a recent culmination in Peterzil and Starchenko [46]. See also the survey paper [21] for a description of the various roles of model theory in the study of o-minimal structures. Chang & Keisler [9] is a basic reference for model theory.

An important event was Wilkie's proof in 1991 that the real exponential field is model-complete and o-minimal, see [64]. This proof cleverly used model-theoretic properties of o-minimal structures, and in return gave new insight into how to establish model-completeness and o-minimality of certain structures on the real field. A theorem of Khovanskii [32] on zerosets of exponential polynomials also plays a key role in [64]. Khovanskii's book [33] elaborates on his theorem and contains interesting applications and open problems.

For other o-minimal structures on the real field, see Van den Dries, Macintyre and Marker [22] and Miller [44]. Building o-minimal structures on the real field has become recently (1996-1997) a very lively activity: among the kinds of functions that turn out to generate o-minimal structures are functions defined by Dirichlet series and by multisummable power series, and Pfaffian functions. (I omit references since at the time of writing much of this is only available in the form of preprints and the dust has not settled yet.)

The following information came too late to include in the references at the end of the book:

1. "Esquisse d'un Programme" has finally appeared (with English translation) in

Geometric Galois Actions. 1. Around Grothendieck's Esquisse d'un Programme, edited by L. Schneps and P. Lochak, Cambridge University Press, 1997. (Teissier's article there discusses tame topology.)

2. Tame topology (in the real setting) is also developed extensively in the book

M. Shiota, *Geometry of subanalytic and semialgebraic sets*, Birkhäuser, Boston, Mass., 1997.

CHAPTER 1

SOME ELEMENTARY RESULTS

Introduction

The seven sections of this chapter are short and quite elementary. Sections 1, 2, and 5 consist of generalities of a purely logical nature that are used throughout this book. In Section 3 we formally introduce o-minimal structures and prove some simple facts about them. In Section 4 we show that o-minimal ordered groups are abelian and divisible, and that o-minimal ordered rings are real closed fields. In Section 6 we prove the o-minimality of dense linearly ordered sets without endpoints via a characterization of their "definable" sets. In Section 7 we do the same for ordered vector spaces over ordered fields, and consider semilinear sets.

§1. Remarks on logical notation and boolean algebras

(1.1) In this volume we will use logical formulas only in an informal and familiar way, as convenient descriptions or definitions of sets and functions.

(1.2) To illustrate this notational use of formulas, let x, y, z be variables ranging over nonempty sets X, Y, Z respectively, and let $\phi(x, y, z)$ and $\psi(x, y, z)$ be conditions on an arbitrary point $(x, y, z) \in X \times Y \times Z$ defining sets

$$\Phi \subseteq X \times Y \times Z \text{ and } \Psi \subseteq X \times Y \times Z,$$
$$\Phi := \{(x, y, z) \in X \times Y \times Z : \phi(x, y, z) \text{ holds}\},$$
$$\Psi := \{(x, y, z) \in X \times Y \times Z : \psi(x, y, z) \text{ holds}\}.$$

Then from the formulas $\phi(x, y, z)$ and $\psi(x, y, z)$ we can construct new formulas:

$\phi(x, y, z) \vee \psi(x, y, z)$, the **disjunction** of $\phi(x, y, z)$ and $\psi(x, y, z)$, defines $\Phi \cup \Psi$;

$\phi(x, y, z) \wedge \psi(x, y, z)$, the **conjunction** of $\phi(x, y, z)$ and $\psi(x, y, z)$, defines $\Phi \cap \Psi$;

$\neg\phi(x, y, z)$, the **negation** of $\phi(x, y, z)$, defines the complement $(X \times Y \times Z) - \Phi$;

$\exists z \phi(x, y, z)$, the **existential quantification** over z of $\phi(x, y, z)$, defines the set

$$\{(x, y) \in X \times Y : \text{ there exists } z \in Z \text{ such that } \phi(x, y, z) \text{ holds}\},$$

in other words, $\exists z\phi(x,y,z)$ defines the set $\pi_Z(\Phi)$, where $\pi_Z: X\times Y\times Z \to X\times Y$ is the obvious projection map; similarly, the formulas $\exists x\phi(x,y,z)$ and $\exists y\phi(x,y,z)$ define the sets $\pi_X(\Phi)$ and $\pi_Y(\Phi)$ where $\pi_X: X\times Y\times Z \to Y\times Z$ and $\pi_Y: X\times Y\times Z \to X\times Z$ are projection maps;

$\forall z\phi(x,y,z)$, the **universal quantification** over z of $\phi(x,y,z)$, defines the set

$$\{(x,y)\in X\times Y: \text{ for all } z\in Z \text{ the condition } \phi(x,y,z) \text{ holds}\}.$$

Note that this is also the set defined by the formula $\neg\exists z\neg\phi(x,y,z)$.

We write $\phi(x,y,z)\to\psi(x,y,z)$ as abbreviation for $(\neg\phi(x,y,z))\vee\psi(x,y,z)$, and $\phi(x,y,z)\leftrightarrow\psi(x,y,z)$ to abbreviate $(\phi(x,y,z)\to\psi(x,y,z))\wedge(\psi(x,y,z)\to\phi(x,y,z))$. Instead of the symbol "$\wedge$" we also use the equivalent symbol "&".

As an example, the formula $\forall y(\exists z\phi(x,y,z)\to\exists z\psi(x,y,z))$ defines the set

$$\{x\in X: \text{ for all } y\in Y \text{ such that there exists } z\in Z \text{ with } \phi(x,y,z), \text{ there exists } z'\in Z \text{ with } \psi(x,y,z')\}.$$

(1.3) Of course, there is nothing special about the use of just three "underlying" sets X, Y, Z. We may as well have more than three, or fewer, and it can happen that $X = Y$, or $Z = X^4$, et cetera. Usually the variables range over the same (nonempty) set, which is often clear from the context and not explicitly specified. The main thing to keep in mind is that the various logical operations on formulas correspond to operations on the sets these formulas define. We shall see in the next two sections that logical notation is more suggestive and transparent than traditional set-theoretic notation, in particular when quantifiers are involved.

(1.4) DEFINITION. A **boolean algebra of subsets of a set** X is a nonempty collection $\mathcal{C}$ of subsets of X such that if $A, B\in\mathcal{C}$, then $A\cup B\in\mathcal{C}$ and $X - A\in\mathcal{C}$.

Note that then $X\in\mathcal{C}$ and $\emptyset\in\mathcal{C}$, and that $A, B\in\mathcal{C}$ implies $A\cap B\in\mathcal{C}$.

An **atom** of a boolean algebra $\mathcal{C}$ of subsets of X is a minimal element (with respect to inclusion) of $\{A\in\mathcal{C}: A\neq\emptyset\}$. For example, the atoms of the boolean algebra $\mathcal{P}(X)$ are exactly the singletons $\{x\}$ with $x\in X$. Given sets $A_1,\ldots,A_n\subseteq X$, we denote by $B(A_1,\ldots,A_n)$ the boolean algebra of subsets of X generated by $A_1,\ldots,A_n$, that is, the smallest boolean algebra of subsets of X that contains $A_1,\ldots,A_n$ as members. (The notation $B(A_1,\ldots,A_n;X)$ would be more correct since it indicates the ambient set X, but X will always be clear from the context.) It is easy to see that the members of $B(A_1,\ldots,A_n)$ are exactly the finite unions of sets of the form

$$(*)\qquad \left(\bigcap_{i\in\Delta} A_i\right)\cap\left(\bigcap_{j\notin\Delta}(X - A_j)\right), \text{ with } \Delta\subseteq\{1,\ldots,n\}.$$

Clearly the nonempty sets of the form $(*)$ are the atoms of $B(A_1,\ldots,A_n)$, so $B(A_1,\ldots,A_n)$ has at most 2^n atoms, and hence at most 2^{2^n} members.

§2. Elementary facts on structures

(2.1) We define a **structure** on a nonempty set R to be a sequence $\mathcal{S} = (\mathcal{S}_m)_{m\in\mathbb{N}}$ such that for each $m \geq 0$:

(S1) $\mathcal{S}_m$ is a boolean algebra of subsets of R^m;
(S2) if $A \in \mathcal{S}_m$, then $R \times A$ and $A \times R$ belong to $\mathcal{S}_{m+1}$;
(S3) $\{(x_1,\ldots,x_m)\in R^m : x_1 = x_m\} \in \mathcal{S}_m$;
(S4) if $A \in \mathcal{S}_{m+1}$, then $\pi(A) \in \mathcal{S}_m$, where $\pi : R^{m+1} \to R^m$ is the projection map on the first m coordinates.

Instead of saying that $\mathcal{S}$ is a structure on R we also say that $(R, \mathcal{S})$ is a structure.

The class of semialgebraic sets mentioned in the Introduction is an example of a structure on $\mathbb{R}$, as will be proved in the next chapter.

Here is another important case. Let Ω be an algebraically closed field, for example, $\Omega = \mathbb{C}$, and let K be a subfield. The **K-constructible subsets of Ω^m** are the finite unions of sets of the form $\{x \in \Omega^m : f_1(x) = \cdots = f_k(x) = 0,\ g(x) \neq 0\}$, where $f_1(X),\ldots,f_k(X), g(X) \in K[X]$ and $X = (X_1,\ldots,X_m)$. That the K-constructible sets form a structure on Ω is essentially Chevalley's constructibility theorem, also known as quantifier elimination for algebraically closed fields. (In the next volume we can give a very quick proof of this theorem. In this book we only use it as illustration.)

The difficulty in proving in these cases that we are dealing with a structure is the verification of axiom (S4): closure under projections. In some other cases $\mathcal{S}$ is defined in such a way that this axiom is trivially satisfied, but then the difficulty is usually located in verifying closure under complements. (In the next volume we will discuss some general logical ideas—quantifier elimination and model completeness—that are often helpful in dealing with such problems.)

In the following lemmas we collect some elementary facts about a structure $\mathcal{S} = (\mathcal{S}_m)$ on R. First, we fix some terminology: a set $A \subseteq R^m$ is said to **belong to** $\mathcal{S}$ (or to **be in** $\mathcal{S}$) if it belongs to $\mathcal{S}_m$; a map $f : A \to B$ with $A \subseteq R^m$ and $B \subseteq R^n$ is said to belong to $\mathcal{S}$ if its graph $\Gamma(f) \subseteq R^{m+n}$ belongs to $\mathcal{S}_{m+n}$.

(2.2) LEMMA.

(i) *If $A \in \mathcal{S}_m$ and $B \in \mathcal{S}_n$, then $A \times B \in \mathcal{S}_{m+n}$.*
(ii) *For $1 \leq i < j \leq m$ the diagonal $\Delta_{ij} := \{(x_1,\ldots,x_m) \in R^m : x_i = x_j\}$ belongs to $\mathcal{S}$.*
(iii) *Let $B \in \mathcal{S}_n$, and let $i(1),\ldots,i(n) \in \{1,\ldots,m\}$. Then the set $A \subseteq R^m$ defined by the condition*

$$(x_1,\ldots,x_m) \in A \Leftrightarrow (x_{i(1)},\ldots,x_{i(n)}) \in B$$

belongs to $\mathcal{S}$. ("Permuting and identifying variables are allowed".)

PROOF. For (i), note that $A \times B = (A \times R^n) \cap (R^m \times B)$, and use (S1) and (S2). For (ii), let $\Delta := \{(x_1, \ldots, x_{j-i}) \in R^{j-i} : x_1 = x_{j-i}\}$, and note that $\Delta_{ij} = R^{i-1} \times \Delta \times R^{m-j+1}$, and use (S3) and (i). For (iii) we note that

$$(x_1, \ldots, x_m) \in A \Leftrightarrow \exists y_1 \ldots \exists y_n \big(x_{i(1)} = y_1 \,\&\, \ldots \,\&\, x_{i(n)} = y_n \,\&\, (y_1, \ldots, y_n) \in B\big).$$

Now think of $(x, y) = (x_1, \ldots, x_m, y_1, \ldots, y_n)$ as ranging over $R^m \times R^n = R^{m+n}$; the formula "$(y_1, \ldots, y_n) \in B$" viewed as a condition on (x, y) defines the set $R^m \times B$ in $\mathcal{S}$, hence by (ii) the formula that follows the quantifiers $\exists y_1 \ldots \exists y_n$ defines a set in $\mathcal{S}$. Applying the quantifiers means taking successive images under the projection maps

$$R^{m+n} \to R^{m+n-1} \to \cdots \to R^m,$$

hence $A \in \mathcal{S}_m$ by (S4). □

(2.3) LEMMA. *Let $S \subseteq R^m$ and let $f : S \to R^n$ be a map that belongs to $\mathcal{S}$, that is, $\Gamma(f) \in \mathcal{S}_{m+n}$. Then we have the following properties:*

(i) *$S \in \mathcal{S}_m$,*
(ii) *if $A \subseteq S$, $A \in \mathcal{S}_m$, then $f(A) \in \mathcal{S}_n$ and the restriction $f|A$ belongs to $\mathcal{S}$,*
(iii) *if $B \in \mathcal{S}_n$, then $f^{-1}(B) \in \mathcal{S}_m$,*
(iv) *if f is injective, its inverse f^{-1} belongs to $\mathcal{S}$,*
(v) *if $f(S) \subseteq T \subseteq R^n$, and $g : T \to R^p$ is a second map belonging to $\mathcal{S}$, then the composition $g \circ f : S \to R^p$ belongs to $\mathcal{S}$.*

PROOF. Let x range over R^m and y over R^n. For (i), use the equivalence $x \in S \Leftrightarrow \exists y (x, y) \in \Gamma(f)$, for (ii) use the equivalence $y \in f(A) \Leftrightarrow \exists x \big(x \in A \,\&\, (x, y) \in \Gamma(f)\big)$, for (iii) use

$$x \in f^{-1}(B) \Leftrightarrow \exists y \big(y \in B \,\&\, (x, y) \in \Gamma(f)\big),$$

for (iv), use part (iii) of the previous lemma, and the equivalence

$$(y, x) \in \Gamma\left(f^{-1}\right) \Leftrightarrow (x, y) \in \Gamma(f).$$

For (v), use the previous lemma, and the equivalence (with z ranging over R^p)

$$(x, z) \in \Gamma(g \circ f) \Leftrightarrow \exists y \big((x, y) \in \Gamma(f) \,\&\, (y, z) \in \Gamma(g)\big). \quad □$$

I hope these short arguments illustrate the utility of logical notation and its interpretation in terms of set operations. There are numerous small facts of this kind that the reader can verify instantaneously, by just writing down a suitable logical formula. Below we give some exercises to become familiar with this technique, which is practiced frequently but often implicitly in these notes. The results of these exercises are also used in the rest of the book.

(2.4) Let R be a nonempty set. Given two structures $\mathcal{S}(1)$ and $\mathcal{S}(2)$ on R we write $\mathcal{S}(1) \subseteq \mathcal{S}(2)$ if $\mathcal{S}(1)_m \subseteq \mathcal{S}(2)_m$ for all m; this defines a partial ordering on the collection of structures on R. Any family $(\mathcal{S}(i))_{i\in I}$ of structures on R has a greatest lower bound $\mathcal{S}$ in the collection of structures on R, namely

$$\mathcal{S} = \bigcap_i \mathcal{S}(i) \text{ with } \mathcal{S}_m := \bigcap_i \mathcal{S}(i)_m, \text{ for each } m.$$

(**2.5**) EXERCISES.

1. Let $A \subseteq R^m$ and let $f = (f_1, \ldots, f_n) : A \to R^n$ be a map with component functions $f_i : A \to R$. Show that f belongs to $\mathcal{S}$ if and only if each f_i belongs to $\mathcal{S}$.

2. (Sheaf property) Let I be a finite index set, let $A \in \mathcal{S}_m$ be the union of the sets $A_i \in \mathcal{S}_m$ $(i \in I)$. Show that a map $f : A \to R^n$ belongs to $\mathcal{S}$ if and only if all its restrictions $f|A_i$ belong to $\mathcal{S}$.

3. Given $A \subseteq R^{m+n}$ and $x \in R^m$ we put $A_x := \{y \in R^n : (x,y) \in A\}$. Show that if $A \in \mathcal{S}_{m+n}$ and $k \in \mathbf{N}$, then the sets $\{x \in R^m : |A_x| \leq k\}$ and $\{x \in R^m : |A_k| = k\}$ belong to $\mathcal{S}_m$.

4. Let the sets A, B, C and the function $f : A \times B \to C$ belong to $\mathcal{S}$. Show that the set $\{a \in A : f(a,\cdot) : B \to C \text{ is injective}\}$ belongs to $\mathcal{S}$, and show also that the set $\{a \in A : f(a,\cdot) : B \to C \text{ is surjective}\}$ belongs to $\mathcal{S}$.

5. Let $P \subseteq R$ be a nonempty subset belonging to $\mathcal{S}_1$. For $m = 0, 1, 2, \ldots$ put

$$(\mathcal{S}|P)_m := \{A \cap P^m : A \in \mathcal{S}_m\}, \text{ a boolean algebra of subsets of } P^m.$$

Show that $\mathcal{S}|P := \big((\mathcal{S}|P)_m\big)_{m \in \mathbf{N}}$ is a structure on P. (The "restriction of $\mathcal{S}$ to P".)

6. Suppose $\mathcal{S}$ contains binary operations $+ : R^2 \to R$ and $\cdot : R^2 \to R$ with respect to which R is a ring (always associative with 1 in this book). Show that $\mathcal{S}$ contains $\{0\}$ and $\{1\}$, and that if $\mathcal{S}$ contains $A \subseteq R^m$ and the functions $f, g : A \to R$, then it contains the functions $-f$, $f + g$, and $f \cdot g$ from A into R.

7. Suppose $R = \mathbb{R}$ and $\mathcal{S}$ contains the order relation $\{(x,y) \in \mathbb{R}^2 : x < y\}$. Show that the topological closure $\mathrm{cl}(A)$ of a set $A \in \mathcal{S}_m$ also belongs to $\mathcal{S}$. Show that if a function $f : \mathbb{R}^{m+1} \to \mathbb{R}$ belongs to $\mathcal{S}$, then the set

$$A := \{a \in \mathbb{R}^m : f(a,t) \text{ tends to a limit } l(a) \in \mathbb{R} \text{ as } t \to +\infty\}$$

belongs to $\mathcal{S}$, and the limit function $l : A \to \mathbb{R}$ so defined belongs to $\mathcal{S}$.

8. Suppose $R = \mathbb{R}$ and $\mathcal{S}$ contains the graphs of addition and multiplication. Show that $\mathcal{S}$ contains the order relation $\{(x,y) \in \mathbb{R}^2 : x < y\}$, and each singleton $\{q\}$ with q a rational number. Show that if $\mathcal{S}$ contains a function $f : I \to \mathbb{R}$, with open $I \subseteq \mathbb{R}$, then it contains the set $I' := \{x \in I : f \text{ is differentiable at } x\}$, and the derivative $f' : I' \to \mathbb{R}$.

(2.6) DIGRESSION: THE SMALLEST STRUCTURE ON THE REAL FIELD CONTAINING $\mathbb{Z}$. In the introduction to this book I mentioned the possibility that, starting with some relatively simple sets in euclidean space and performing the usual elementary operations on them, we obtain ever more complicated sets when we iterate this generation process indefinitely. Here is the main example of this phenomenon.

Let $\mathcal{S}$ be the smallest structure on the set $\mathbb{R}$ that contains the graphs of addition and multiplication on $\mathbb{R}$, the subset $\mathbb{Z}$ of $\mathbb{R}$, and each singleton $\{r\}$, $r \in \mathbb{R}$. Then each $\mathcal{S}_n$ consists exactly of the so-called projective subsets of $\mathbb{R}^n$, in particular, all Borel sets in $\mathbb{R}^n$ belong to $\mathcal{S}_n$; see for example Exercise 37.6 in the book *Classical descriptive set theory* by A. Kechris (Springer-Verlag, 1995).

The set $\mathbb{Z}$ of integers can of course be replaced here by the sine function on $\mathbb{R}$, since $\mathbb{Z} = \{a \in \mathbb{R} : \sin(\pi a) = 0\}$.

Those familiar with descriptive set theory and with Matijasevich's theorem on diophantine sets may find it an amusing exercise to show that one can obtain all open subsets of $\mathbb{R}^n$ from a single polynomial equation over $\mathbb{Z}$:

There is a polynomial $P_n(X, Y, Z)$ with integer coefficients in the variables $X = (X_0, X_1, \ldots, X_n)$, $Y = (Y_1, \ldots, Y_p)$, $Z = (Z_1, \ldots, Z_q)$ (for certain $p, q \in \mathbb{N}$ depending on n) such that $\{A_r : r \in \mathbb{R}\}$ is equal to the collection of all open subsets of $\mathbb{R}^n$, where $A \subseteq \mathbb{R}^{n+1}$ is the image of the set

$$\{(x, y, z) \in \mathbb{R}^{n+1+p} \times \mathbb{Z}^q : P_n(x, y, z) = 0\}$$

under the projection map $\mathbb{R}^{n+1+p+q} \to \mathbb{R}^{n+1}$ onto the first $n + 1$ coordinates.

Once all open sets in euclidean spaces are available we also obtain their complements, the closed sets, hence all real-valued continuous functions on closed sets (which have closed graphs), and so on. See Kechris's book referred to earlier for how this generation process continues and never stops.

§3. O-minimal structures

(3.1) NOTATIONS AND CONVENTIONS.

Given linearly ordered sets R_1 and R_2 and a map $f : R_1 \to R_2$ we say:

f is **strictly increasing** if $x < y$ in R_1 implies $f(x) < f(y)$ in R_2,
increasing if $x \leq y$ in R_1 implies $f(x) \leq f(y)$ in R_2,
strictly decreasing if $x < y$ in R_1 implies $f(x) > f(y)$ in R_2,
decreasing if $x \leq y$ in R_1 implies $f(x) \geq f(y)$ in R_2,
strictly monotone if f is either strictly increasing or strictly decreasing,
monotone if f is either increasing or decreasing.

A linearly ordered set R is called **dense** if for all $a, b \in R$ with $a < b$ there is $c \in R$ with $a < c < b$. A subset X of a linearly ordered set R is called **convex** (in R) if $a < b < c$ with $a, c \in X$ implies $b \in X$.

Let $(R, <)$ be a dense linearly ordered nonempty set without endpoints. (That $(R, <)$ has no endpoints means that R has no largest or smallest element.)

For convenience, we add two endpoints $-\infty$ and $+\infty$, with $-\infty < a < +\infty$ for all $a \in R$, and put $R_\infty := R \cup \{-\infty, +\infty\}$. An **interval** is always a nonempty "open" interval

$$(a, b) := \{x \in R : a < x < b\} \text{ with } -\infty \leq a < b \leq +\infty.$$

Note that intervals are convex.

Here are some further notations for certain kinds of convex subsets of R:

$$\begin{aligned}
(a, b] &:= \{x \in R : a < x \leq b\} \text{ where } -\infty \leq a < b < +\infty, \\
[a, b) &:= \{x \in R : a \leq x < b\} \text{ where } -\infty < a < b \leq +\infty, \\
[a, b] &:= \{x \in R : a \leq x \leq b\} \text{ where } -\infty < a \leq b < +\infty.
\end{aligned}$$

(N.B. Sets of these three kinds will never be referred to as intervals in this book.)

REMARK. We use the same notation for an interval $(a, b) \subseteq R$ as for an ordered pair $(a, b) \in R^2$. It should always be clear from the context which is meant.

We equip R with the **interval topology** (the intervals form a base), and each product R^m with the corresponding product topology, a base of which is formed by the boxes in R^m: a **box in** R^m is a cartesian product $(a_1, b_1) \times \cdots \times (a_m, b_m)$ of intervals. Note that R^m is a Hausdorff space with this topology.

The (topological) closure in R^m of a set $A \subseteq R^m$ is denoted by $\mathrm{cl}(A)$, and its interior (in R^m) by $\mathrm{int}(A)$.

Given functions $f, g : X \to R_\infty$ on a set $X \subseteq R^m$ we put

$$\begin{aligned}
(f, g) &:= \{(x, r) \in X \times R : f(x) < r < g(x)\}, \\
[f, g] &:= \{(x, r) \in X \times R_\infty : f(x) \leq r \leq g(x)\}.
\end{aligned}$$

We consider (f, g) as a subset of R^{m+1}; also $[f, g] \subseteq R^{m+1}$ if f and g are R-valued.

We write $f < g$ to indicate that $f(x) < g(x)$ for all $x \in X$.

(3.2) Let $(R, <)$ be a dense linearly ordered nonempty set without endpoints. An **o-minimal structure on** $(R, <)$ is by definition a structure $\mathcal{S}$ on R such that

(O1) $\{(x, y) \in R^2 : x < y\} \in \mathcal{S}_2$,
(O2) the sets in $\mathcal{S}_1$ are exactly the finite unions of intervals and points.

In that case we also say that $(R,<,\mathcal{S})$ is an o-minimal structure.

For the rest of this section we fix an o-minimal structure $\mathcal{S}$ on $(R,<)$.

CONVENTION. Instead of saying that a set A belongs to $\mathcal{S}$ we usually follow more traditional terminology and say that A is **definable**, and similarly with maps. This is of course only permitted if it is clear from the context which $\mathcal{S}$ is meant.

The lemmas that follow are quite simple and will be used frequently without explicit reference.

(3.3) LEMMA. *Let $A \subseteq R$ be definable. Then:*

(i) $\inf(A)$ *and* $\sup(A)$ *exist in* R_∞ *(Dedekind completeness for definable sets),*
(ii) *the boundary* $bd(A) := \{x \in R :$ each interval containing x intersects both A and $R - A\}$ *is finite, and if* $a_1 < \cdots < a_k$ *are the points of* $\mathrm{bd}(A)$ *in order, then each interval* (a_i, a_{i+1})*, where* $a_0 = -\infty$ *and* $a_{k+1} = +\infty$*, is either part of A or disjoint from A.*

This is almost immediate from the definition of o-minimality. Here are other easy results, the first one a variant of exercise 7 in (2.5) above.

(3.4) LEMMA.

(i) *If $A \subseteq R^m$ is definable, so are* $\mathrm{cl}(A)$ *and* $\mathrm{int}(A)$.
(ii) *If $A \subseteq B \subseteq R^m$ are definable sets, and A is open in B, then there is a definable open $U \subseteq R^m$ with $U \cap B = A$.*

PROOF. For (i), note the following equivalence:

$$(x_1,\ldots,x_m) \in \mathrm{cl}(A)$$

$$\Longleftrightarrow$$

$$\forall y_1 \ldots \forall y_m \forall z_1 \ldots \forall z_m \big[(y_1 < x_1 < z_1 \;\&\; \ldots \;\&\; y_m < x_m < z_m) \rightarrow$$
$$\exists a_1 \ldots \exists a_m \big(y_1 < a_1 < z_1 \;\&\; \ldots \;\&\; y_m < a_m < z_m \;\&\; (a_1,\ldots,a_m) \in A\big)\big].$$

Now use the interpretation of the logical symbols in terms of operations on sets. For (ii), note that one can take for U the union of all boxes in R^m whose intersection with B is contained in A. □

REMARK. Lemma (3.4) uses only axiom (O1) of o-minimality, and not (O2).

Next we introduce a notion of connectedness appropriate for definable sets. In Chapter 3, (2.19), exercise 7 we show that if the underlying ordered set $(R,<)$ of $(R,<,\mathcal{S})$ is the ordered set of real numbers, then this agrees (for definable sets) with the usual notion of connectedness.

(3.5) DEFINITION. A set $X \subseteq R^m$ is called **definably connected** if X is definable and X is not the union of two disjoint nonempty definable open subsets of X.

The verification of the following easy results is left as an exercise.

(3.6) LEMMA.

(1) *The definably connected subsets of R are the following: the empty set, the intervals, the sets $[a, b)$ with $-\infty < a < b \leq +\infty$, the sets $(a, b]$ with $-\infty \leq a < b < +\infty$, and the sets $[a, b]$ with $-\infty < a \leq b < +\infty$.*

(2) *The image of a definably connected set $X \subseteq R^m$ under a definable continuous map $f: X \to R^n$ is definably connected.*

(3) *If X and Y are definable subsets of R^m, $X \subseteq Y \subseteq \text{cl}(X)$, and X is definably connected, then Y is definably connected.*

(4) *If X and Y are definably connected subsets of R^m and $X \cap Y \neq \emptyset$, then $X \cup Y$ is definably connected.*

Note the following special case of (2).

If the function $f: [a, b] \to R$ is definable and continuous, then f assumes all values between $f(a)$ and $f(b)$.

At this point it would be possible to continue directly with Chapters 2, 3, 4, and 5, since logically these depend only on the material just treated.

(3.7) EXERCISE. Let $S \subseteq R^{m+n}$ be definable. Show:

(i) $\{x \in R^m : S_x \text{ is open}\}$ is definable.

(ii) $\{(x, y) \in R^{m+n} : y \in \text{int}(S_x)\}$ is definable.

§4. O-minimal ordered groups and rings

(4.1) To formulate the next result we define an **ordered group** to be a group equipped with a linear order that is invariant under left and right multiplication:

$$x < y \Rightarrow zx < zy \text{ and } xz < yz.$$

(So the additive group of reals, and the multiplicative group of positive reals, are ordered groups under the usual ordering, but the multiplicative group of all nonzero reals is not.)

(4.2) PROPOSITION. *Suppose $(R, <, \mathcal{S})$ is an o-minimal structure and $\mathcal{S}$ contains a binary operation $\cdot$ on R, such that $(R, <, \cdot)$ is an ordered group. Then the group $(R, \cdot)$ is abelian, divisible and torsion-free.*

Denoting the unit element of the group by 1 we first prove

LEMMA. *The only definable subsets of R that are also subgroups are $\{1\}$ and R.*

PROOF. Given a definable subgroup G we first show that G is convex: if not, then there are $g \in G$, $r \in R - G$ with $1 < r < g$. This gives a sequence

$$1 < r < g < rg < g^2 < rg^2 < g^3 < \cdots$$

whose terms alternate in being in and out of the definable set G, which is impossible. So G is convex, hence assuming $G \neq \{1\}$ we have $s := \sup(G) > 1$ with $(1, s) \subseteq G$. If $s = +\infty$, then clearly $G = R$. If $s < +\infty$, then we take any $g \in (1, s)$, and obtain $s = g \cdot g^{-1}s \in G$, since $g^{-1}s \in (1, s)$, hence $s < gs \in G$, contradicting the definition of s.

PROOF OF PROPOSITION (4.2). For each $r \in R$ the centralizer $C_r := \{x \in R : rx = xr\}$ is a definable subgroup containing r, so $C_r = R$ by the lemma. Hence R is abelian. For each $n > 0$ the subgroup $\{x^n : x \in R\}$ is definable, hence equal to R. This gives divisibility. Every ordered group is torsion-free. □

(4.3) REMARK. Let $(R, <, +)$ be an ordered abelian group, $R \neq \{0\}$, so $(R, <)$ has no endpoints. Assume also that the linearly ordered set $(R, <)$ is dense. Then the addition operation $+ : R^2 \to R$ and the additive inverse operation $- : R \to R$ are continuous with respect to the interval topology, that is, $(R, +)$ is a topological group with respect to the interval topology.

(4.4) In this book an **ordered ring** is a ring (associative with 1) equipped with a linear order $<$ such that

(i) $0 < 1$;
(ii) $<$ is translation invariant, that is: $x < y \Rightarrow x + z < y + z$;
(iii) $<$ is invariant under multiplication by positive elements:

$$(x < y \text{ and } z > 0 \Rightarrow xz < yz).$$

Note that then the additive group of the ring is an ordered group in the previous sense, that the ring has no zero divisors, that $x^2 \geq 0$ for all x, and that

$$k \mapsto k \cdot 1 \colon \mathbf{Z} \to \text{ring}$$

is a strictly increasing ring embedding, with the usual ordering on the ring $\mathbf{Z}$ of integers; we shall always identify the ordered ring $\mathbf{Z}$ with its image in the ring under this embedding.

Suppose our ordered ring is moreover a **division ring**: for each $x \neq 0$ there is y with $x \cdot y = 1$. It is easy to check that such a y is unique (written as x^{-1}), and satisfies $y \cdot x = 1$, and that $x > 0$ implies $y > 0$. It is also easy to see that the additive group is divisible, the underlying ordered set is dense without endpoints, and the maps $(x, y) \mapsto xy$ and $x \mapsto x^{-1}$ (the last map defined only for $x \neq 0$) are continuous with respect to the interval topology.

(4.5) An **ordered field** is an ordered division ring with commutative multiplication. Examples of ordered fields are the field of reals and the field of rational numbers with the usual ordering. In this book we define a **real closed field** to be an ordered field such that if $f(X)$ is a one-variable polynomial with coefficients in the field and $a < b$ are elements in the field with $f(a) < 0 < f(b)$, then there is $c \in (a, b)$ in the field with $f(c) = 0$. (Intermediate value property.) So the ordered field of reals is real closed and the ordered field of rational numbers is not real closed. In the next volume we develop enough model theory so that we can give an efficient and brief treatment of the theory of real closed fields of Artin and Schreier, including Tarski's elimination theory (and much more). Here we will restrict ourselves to the following:

(4.6) PROPOSITION. *Suppose $(R, <, \mathcal{S})$ is an o-minimal structure and $\mathcal{S}$ contains binary operations $+ : R^2 \to R$ and $\cdot : R^2 \to R$ such that $(R, <, +, \cdot)$ is an ordered ring. Then $(R, <, +, \cdot)$ is a real closed field.*

PROOF. For each $r \in R$ we have a definable additive subgroup rR of $(R, +)$, hence $rR = R$ if $r \neq 0$, by the previous proposition. This shows that $(R, <, +, \cdot)$ is an ordered division ring. Let $\mathrm{Pos}(R) := \{r \in R : r > 0\}$. Clearly $\mathrm{Pos}(R)$ is an ordered multiplicative group. By restricting $\mathcal{S}$ to $\mathrm{Pos}(R)$ (see (2.5), exercise 5) it follows from the previous proposition that multiplication is commutative on $\mathrm{Pos}(R)$, hence on all of R. So $(R, <, +, \cdot)$ is an ordered field. By exercise 6 of (2.5), and the remarks in (4.3) and (4.4), each one-variable polynomial $f(X) \in R[X]$ gives rise to a definable continuous function $x \mapsto f(x) : R \to R$. Now apply the remark at the end of (3.6) to this function. □

§5. Model-theoretic structures

(5.1) A structure $\mathcal{S}$ on a set R is often introduced by indicating relations and functions on R that "generate" $\mathcal{S}$. For example, in the next chapter we will see that the structure on $\mathbb{R}$ generated by its addition and multiplication operations, and the individual real numbers, consists exactly of the semialgebraic sets. It is quite useful also for theoretical reasons to have available a notion of structure that singles out particular relations and functions as "primitives".

(5.2) DEFINITION. A **model-theoretic structure** $\mathcal{R} = (R, (S_i)_{i\in I}, (f_j)_{j\in J})$ consists of a nonempty set R, relations $S_i \subseteq R^{m(i)}$ ($i \in I, m(i) \in \mathbb{N}$), and functions $f_j : R^{n(j)} \to R$ ($j \in J, n(j) \in \mathbb{N}$). If $n(j) = 0$, we identify f_j with its unique value in R, and call f_j a **constant**. We also call R the **underlying set** of $\mathcal{R}$, and the S_i and f_j the **basic** (or **primitive**) relations and functions of $\mathcal{R}$.

If the index sets I and J are finite we usually just list the relations and functions.

For example, an abelian group will be considered as a model-theoretic structure of the form $(A, 0, -, +)$, with 0 the zero element, $- : A \to A$ the operation of taking the negative and $+ : A^2 \to A$ the group operation; an ordered ring is a model-theoretic structure of the form $(R, <, 0, 1, -, +, \cdot)$, et cetera.

Given a model-theoretic structure $\mathcal{R} = \big(R, (S_i)_{i \in I}, (f_j)_{j \in J}\big)$, and set $C \subseteq R$ we define a new model-theoretic structure $\mathcal{R}_C := \big(R, (S_i)_{i \in I}, (f_j)_{j \in J}, (c)_{c \in C}\big)$ which has the same underlying set and the same basic relations as $\mathcal{R}$, but has in addition to the basic functions f_j a constant c for each element $c \in C$, where formally we identify c with the corresponding function $R^0 \to R$.

(5.3) Given a model-theoretic structure $\mathcal{R} = \big(R, (S_i), (f_j)\big)$, we let $\mathrm{Def}(\mathcal{R})$ be the smallest structure on the set R (in the sense of Section 2) that contains each relation S_i and each function f_j. Note that for each set $C \subseteq R$ the structure $\mathrm{Def}(\mathcal{R}_C)$ on R is larger than or equal to $\mathrm{Def}(\mathcal{R})$. (We can recover $\mathrm{Def}(\mathcal{R}_C)$ from $\mathrm{Def}(\mathcal{R})$ and C in a simple way, see (5.9), exercise 1 below.) In this book we mostly work with $\mathrm{Def}(\mathcal{R}_R)$. The sets $S \subseteq R^m$ and maps $f : S \to R^n$ that belong to $\mathrm{Def}(\mathcal{R})$ are also called **definable in** $\mathcal{R}$ (or just definable, when $\mathcal{R}$ is clear from the context). A point $a = (a_1, \ldots, a_m) \in R^m$ is said to be **definable in** $\mathcal{R}$ if the set $\{a\} \subseteq R^m$ is definable in $\mathcal{R}$.

We usually write "definable in $\mathcal{R}$ using constants from C", or "definable in $\mathcal{R}$ with parameters from C" in place of "definable in $\mathcal{R}_C$". For $C = R$ we write "definable in $\mathcal{R}$ using constants" or "definable in $\mathcal{R}$ with parameters" instead of "definable in $\mathcal{R}_R$". We often omit "in $\mathcal{R}$", if $\mathcal{R}$ is clear from the context.

(5.4) To discuss in a uniform way all model-theoretic structures of a certain type, for instance all ordered abelian groups, or all real closed fields, it is convenient to have available the notions of a (first-order) language L and L-structure.

A **language** L is a disjoint union of two sets, a set of **relation symbols**, and a set of **function symbols**, each relation symbol S and each function symbol f being equipped with a number: $\mathrm{arity}(S), \mathrm{arity}(f) \in \mathbf{N}$. If $\mathrm{arity}(S) = m, \mathrm{arity}(f) = n$, we also say that S is an m-ary relation symbol, and f is an n-ary function symbol. Function symbols of arity 0 are also called **constant symbols**, and instead of saying "1-ary" and "2-ary" we say "unary" and "binary".

(5.5) An L-**structure** is a model-theoretic structure $\mathcal{R} = \big(R, (S^{\mathcal{R}}), (f^{\mathcal{R}})\big)$ such that for each m-ary relation symbol S and each n-ary function symbol f of L we have

$$S^{\mathcal{R}} \subseteq R^m, \quad f^{\mathcal{R}} : R^n \to R;$$

$S^{\mathcal{R}}$ and $f^{\mathcal{R}}$ are the **interpretations** of the symbols S and f in $\mathcal{R}$. So the set of relation symbols of L indexes the primitive relations of $\mathcal{R}$, and the set of function symbols of L indexes the primitive functions of $\mathcal{R}$. We usually drop the superscript $\mathcal{R}$ in $S^{\mathcal{R}}$ and $f^{\mathcal{R}}$, leaving it to the context whether the symbol or its interpretation is meant.

(5.6) EXAMPLES.

(1) The language of abelian groups is $L_{ab} := \{0, -, +\}$: it has a constant symbol 0, unary function symbol $-$, and binary function symbol $+$. We consider abelian groups as L_{ab}-structures in the obvious way.

(2) The language $L_{ab}(<) := \{<, 0, -, +\}$ of ordered abelian groups has in addition a binary relation symbol $<$, and we regard ordered abelian groups as $L_{ab}(<)$-structures in the obvious way.

(3) The language of rings is $L_{ring} := \{0, 1, -, +, \cdot\}$: it extends the language of abelian groups by an extra constant symbol 1 and an extra binary function symbol $\cdot$. We consider rings as L_{ring}-structures in the obvious way.

(In a later volume we develop enough model theory to make better use of the notion of L-structure. In this book it is only a matter of convenient terminology.)

(5.7) DEFINITION. A model-theoretic structure $\mathcal{R} = (R, <, \ldots)$, where $<$ is a dense linear order without endpoints on R, is called **o-minimal** if $\mathrm{Def}(\mathcal{R}_R)$ is an o-minimal structure on $(R, <)$, in other words, every set $S \subseteq R$ that is definable in $\mathcal{R}$ using constants is a union of finitely many intervals and points.

(5.8) EXPANSIONS. The simplest thing is just to say how we use this notion in the few places that it occurs in this book. Let $(R, <, \mathcal{S})$ be an o-minimal structure. We say that $(R, <, \mathcal{S})$ **expands an ordered abelian group** if there are abelian group operations $0 : R^0 \to R$, $- : R \to R$, and $+ : R^2 \to R$ belonging to $\mathcal{S}$ such that $(R, <, 0, -, +)$ is an ordered abelian group; in that case we also say that $(R, <, \mathcal{S})$ **expands the ordered abelian group** $(R, <, 0, -, +)$. In particular, the underlying ordered set of $(R, <, \mathcal{S})$ is the same as the underlying ordered set of the ordered abelian group that $(R, <, \mathcal{S})$ expands. Similarly, we define what it means for $(R, <, \mathcal{S})$ to expand an ordered ring. Note that by the previous section, if the o-minimal structure $(R, <, \mathcal{S})$ expands a certain ordered ring, then this ordered ring is actually a real closed field.

(5.9) EXERCISES.

1. Let $\mathcal{R} = (R, \ldots)$ be a model-theoretic structure, $C \subseteq R$, and $A \subseteq R^n$. Show that A is definable in $\mathcal{R}$ using constants from C if and only if there exist a set $S \subseteq R^{m+n}$ definable in $\mathcal{R}$, and elements $c_1, \ldots, c_m \in C$ such that for all $(x_1, \ldots, x_n) \in R^n$

$$(x_1, \ldots, x_n) \in A \Leftrightarrow (c_1, \ldots, c_m, x_1, \ldots, x_n) \in S.$$

2. Let $\mathcal{R} = (R, \ldots)$ be a model-theoretic structure and $S \subseteq R^{m+n}$ definable in $\mathcal{R}$. Show that if $a \in R^m$ is definable in $\mathcal{R}$, then the set $S_a := \{b \in R^n : (a, b) \in S\}$ is definable in $\mathcal{R}$.

3. Let $\mathcal{R} = (\mathbb{R}, <, 0, 1, -, +, \cdot)$ be the ordered field of real numbers. Show that each function $f : \mathbb{R}^m \to \mathbb{R}$ defined by a polynomial $f(X_1, \ldots, X_m) \in \mathbb{R}[X_1, \ldots, X_m]$

is definable in $\mathcal{R}$ using constants. Derive that each semialgebraic set in $\mathbb{R}^m$ (see "Introduction and Overview") is definable in $\mathcal{R}$ using constants.

§6. The simplest o-minimal structures

(6.1) Let $(R,<)$ be a dense linearly ordered nonempty set without endpoints.

We prove below that the model-theoretic structure $(R,<)$ is o-minimal, by describing explicitly all definable sets. These o-minimal structures are too simple to be of much interest, but we treat this case to illustrate the notions of the previous sections. Starting with Chapter 3 we develop systematically the general theory of o-minimal structures. However, in Chapters 6–10 we will assume that our o-minimal structures expand ordered abelian groups or even ordered rings. Such extra assumptions are satisfied in all cases of real interest to us, but not by $(R,<)$, discussed in the present section. In this connection $(R,<)$ can serve as a useful counterexample to potential generalizations of various theorems in these later chapters, showing that the extra assumptions in Chapters 6–10 cannot be omitted.

(6.2) Let $1 \le i \le m$. The function $(x_1,\ldots,x_m) \mapsto x_i : R^m \to R$ will be denoted just by x_i. The **simple** functions on R^m are by definition these coordinate functions $x_1,\ldots,x_m$, and the constant functions $R^m \to R$.

(6.3) Let $f_1,\ldots,f_N$ be simple functions on R^m, and let $\epsilon:\{1,\ldots,N\}^2 \to \{-1,0,1\}$ be given. Then we put

$$\begin{aligned}\epsilon(f_1,\ldots,f_N) := \{x \in R^m : \text{ for all } (i,j) \in \{1,\ldots,N\}^2 \text{ we have }\\ f_i(x) < f_j(x) \text{ if } \epsilon(i,j) = -1,\\ f_i(x) = f_j(x) \text{ if } \epsilon(i,j) = 0,\\ f_i(x) > f_j(x) \text{ if } \epsilon(i,j) = 1\}.\end{aligned}$$

Of course $\epsilon(f_1,\ldots,f_N)$ may be empty, for instance when $\epsilon(i,j) \ne -\epsilon(j,i)$ for some pair (i,j). Suppose that $\epsilon(f_1,\ldots,f_N)$ is nonempty. Note that if ξ and η are the restrictions of f_i and f_j to $\epsilon(f_1,\ldots,f_N)$, then either $\xi < \eta$, or $\xi = \eta$, or $\eta < \xi$. Let $\xi_1 < \cdots < \xi_k$ be the restrictions of $f_1,\ldots,f_N$ to $\epsilon(f_1,\ldots,f_N)$ arranged in increasing order. One checks easily that the sets $\Gamma(\xi_j)$ $(1 \le j \le k)$ and the sets (ξ_j,ξ_{j+1}) $(0 \le j \le k$, where $\xi_0 = -\infty$ and $\xi_{k+1} = +\infty$ by convention) are exactly the nonempty subsets of R^{m+1} of the form $\epsilon'(f_1,\ldots,f_N,x_{m+1})$, where

$$\epsilon':\{1,\ldots,N,N+1\}^2 \to \{-1,0,1\}$$

is an extension of ϵ. (Here a simple function f on R^m is regarded as a simple function on R^{m+1} by setting $f(x_1,\ldots,x_m,x_{m+1}) := f(x_1,\ldots,x_m)$.) Define a **simple set in** R^m to be a subset of R^m of the form $\epsilon(f_1,\ldots,f_N)$ with $f_1,\ldots,f_N$ simple functions on R^m and $\epsilon:\{1,\ldots,N\}^2 \to \{-1,0,1\}$. We have just proved that if $S \subseteq R^{m+1}$ is simple, then its image under the projection map

$$(x_1,\ldots,x_m,x_{m+1}) \mapsto (x_1,\ldots,x_m): R^{m+1} \to R^m$$

is simple in R^m. We now easily derive

(6.4) Proposition. *The subsets of R^m that are definable in $(R,<)$ using constants are exactly the finite unions of simple sets in R^m.*

Proof. Let $\mathcal{S}_m$ be the collection of finite unions of simple sets in R^m. Clearly $\mathcal{S}_m$ is a boolean algebra of subsets of R^m, and each set in $\mathcal{S}_m$ is definable in $(R,<)$ using constants. The considerations in (6.3) above show that $\mathcal{S} := (\mathcal{S}_m)_{m \in \mathbf{N}}$ is a structure on the set R, hence the sets in $\mathcal{S}_m$ are exactly the subsets of R^m definable in $(R,<)$ using constants. □

Remark. It follows easily that $\mathcal{S}$ is an o-minimal structure on $(R,<)$.

(6.5) Corollary. *The model-theoretic structure $(R,<)$ is o-minimal.*

(6.6) From a geometric viewpoint $(R,<)$ can have rather bad properties. For instance, given any cardinal κ one can construct a dense linearly ordered nonempty set $(R,<)$ without endpoints and a collection $\mathcal{C}$ of κ intervals in $(R,<)$ such that $|I| \neq |J|$ for distinct $I, J \in \mathcal{C}$. We leave the construction of such an $(R,<)$ and collection $\mathcal{C}$ of intervals to the reader as an exercise, and note only the extreme lack of homogeneity of such an $(R,<)$ for large κ: $\mathcal{C}$ is a collection of κ intervals no two of which are homeomorphic. In Chapters 6–10 we impose extra algebraic structure that implies homogeneity.

§7. Semilinear sets

(7.1) In this section we show that the sets definable using constants in an ordered vector space over an ordered field are exactly the semilinear sets. Ordered vector spaces occur naturally in a variety of situations, but in this book the material of this section will only play a minor role.

(7.2) Definitions. *Let F denote an ordered field, until further notice.*

An **ordered F-linear space** is a vector space R over F equipped with a linear order making R an ordered additive group, such that for all $\lambda \in F$ and $x \in R$

$$(\lambda > 0, x > 0) \Rightarrow \lambda x > 0.$$

Instead of "ordered F-linear space" we also say "ordered vector space over F". Of course, F considered as a vector space over itself, with its usual order, is an ordered F-linear space. More generally, any ordered field extension of F is an ordered F-linear space in an obvious way. Note also that for $F = \mathbf{Q}$ the ordered $\mathbf{Q}$-linear spaces are exactly the divisible ordered abelian groups, where the latter are made into vector spaces over $\mathbf{Q}$ in the obvious way.

(7.3) *Let R be an ordered F-linear space. We assume $R \neq \{0\}$, so that $(R,<)$ is a dense linearly ordered set without endpoints.*

An **affine function on R^m** is a function $f: R^m \to R$ of the form

$$f(x_1, \ldots, x_m) = \lambda_1 x_1 + \cdots + \lambda_m x_m + a,$$

where the $\lambda_i \in F$ and $a \in R$ are fixed. A **basic semilinear set in R^m** is a set of the form

$$\{x \in R^m : f_1(x) = \cdots = f_p(x) = 0, \ g_1(x) > 0, \ldots, g_q(x) > 0\},$$

with the f_i and g_j affine functions on R^m. (We allow of course $p = 0$ or $q = 0$.) A **semilinear set in R^m is a finite union of basic semilinear sets in R^m**. (Perhaps "semiaffine" would be a better term than "semilinear", but "semilinear" seems more popular.)

Note that the intersection of two basic semilinear sets in R^m is a basic semilinear set in R^m. The basic semilinear sets in $R = R^1$ are exactly the intervals, the points, and the empty set. The complement in R^m of a basic semilinear set in R^m is a semilinear set in R^m, but not necessarily a basic semilinear set in R^m.

(7.4) Proposition. *Let $f_1, \ldots, f_N$ be affine functions on R^{m+1}. Then we can partition R^m into basic semilinear sets $C_1, \ldots, C_k$ such that for each $C \in \{C_1, \ldots, C_k\}$ there are functions $\xi_{C,1} < \cdots < \xi_{C,j(C)}$ from C into R with the following properties:*

(i) *each $\xi_{C,j}$ is the restriction of an affine function on R^m;*
(ii) *each function f_n $(1 \le n \le N)$ has constant sign on each of the sets $\Gamma(\xi_{C,j})$ $(1 \le j \le j(C))$, and on each of the sets $(\xi_{C,j}, \xi_{C,j+1})$ $(0 \le j \le j(C)$, where $\xi_{C,0} := -\infty$, $\xi_{C,j(C)+1} := +\infty$ by convention).*

Proof. Let $f_1, \ldots, f_M$ depend on the last variable x_{m+1}, while the f_n for $M < n \le N$ do not. The latter may therefore be viewed also as functions on R^m. Write $f_p(x_1, \ldots, x_m, x_{m+1}) = \lambda_p(g_p(x_1, \ldots, x_m) - x_{m+1})$ for $1 \le p \le M$, where $\lambda_p \in F^\times$ and g_p is an affine function on R^m. Take a partition of R^m into basic semilinear sets $C_1, \ldots, C_k$ on each of which each function in

$$\{f_{M+1}, \ldots, f_N\} \cup \{g_p - g_q : 1 \le p, \ q \le M\}$$

takes constant sign. For each $C \in \{C_1, \ldots, C_k\}$, let $\xi_{C,1} < \cdots < \xi_{C,j(C)}$ be the restrictions of the functions g_p to C arranged in increasing order. This is possible because the $g_p - g_q$ are of constant sign on C. Then the desired result holds. □

(7.5) Remark. Note that each of the sets in (ii) is itself a basic semilinear set. The actual construction in the proof produces sets as in (ii) that are of the form

$$\{(x,t) \in C \times R : \ \mathrm{sign}(f_n(x,t)) = \epsilon(n) \text{ for } n = 1, \ldots, N\}$$

for a suitable sign description $\epsilon: \{1, \ldots, N\} \to \{-1, 0, 1\}$. This property is somewhat stronger than (ii).

(7.6) Corollary. *Let $\mathcal{S}_m$ be the boolean algebra of semilinear subsets of R^m. Then $\mathcal{S} := (\mathcal{S}_m)_{m \in \mathbb{N}}$ is an o-minimal structure on the ordered set $(R, <)$. Each function in $\mathcal{S}$ is piecewise affine, more precisely, given a function $f: A \to R$ in $\mathcal{S}$ with $A \subseteq R^m$, there is a partition of A into basic semilinear sets A_i $(1 \le i \le k)$ such that $f|A_i$ is the restriction to A_i of an affine function on R^m, for each $i \in \{1, \ldots, k\}$.*

Proof. Let $\pi: R^{m+1} \to R^m$ be the projection map onto the first m coordinates. Given a set $S \in \mathcal{S}_{m+1}$, we have to show that $\pi(S) \in \mathcal{S}_m$. Let $f_1, \ldots, f_N$ be the affine functions involved in a description of S as a union of basic semilinear sets. Applying the proposition above to $f_1, \ldots, f_N$ we see that S is a union of sets as described in (ii). Since each of the sets in (ii) has as its π-image a basic semilinear set in R^m, it follows that $\pi(S) \in \mathcal{S}_m$. This shows that $\mathcal{S}$ is indeed an o-minimal structure on $(R, <)$. If S as above is the graph of a function $f: A \to R$, $A \subseteq R^m$, then the sets of type (ii) that are part of S must be of the form $\Gamma(\xi)$, with ξ a restriction of an affine function. Hence f is piecewise affine. □

(7.7) Construe R as a model-theoretic structure for the language

$$L_F := \{<, 0, -, +\} \cup \{\lambda\cdot : \ \lambda \in F\}$$

of ordered abelian groups augmented by a unary function symbol $\lambda\cdot$ for each $\lambda \in F$, to be interpreted as multiplication by the scalar λ.

(7.8) Corollary. *The subsets of R^m definable in the L_F-structure R using constants are exactly the semilinear sets in R^m.*

Proof. We leave it to the reader to derive this from (7.6). □

(7.9) Remark. The proofs above are "uniform in R", and one particular consequence of this uniformity is that if R' is an ordered F-linear space extending R, then an "elementary statement with constants from R" holds in R if and only if it holds in R'. Since F can be embedded as an ordered F-linear space in each nontrivial ordered F-linear space, it follows that all nontrivial ordered F-linear spaces satisfy exactly the same elementary statements formulated in the language L_F. These issues will be rigorously treated in a later volume where the necessary logical preliminaries are put in place. (For instance, we have to define "elementary statement".) In this book we mainly deal with the situation that $R = F$, and we then refer to the semilinear sets in R^m for $m = 0, 1, 2, \ldots$ as semilinear sets over R.

(7.10) We never used the commutativity of the multiplication on F: everything in this section goes through when F is an ordered division ring. ("Vector space over the division ring F" is then interpreted as "left F-module".) Noncommutative ordered division rings can be obtained from noncommutative ordered groups via a power series construction, see for example Fuchs [25].

(7.11) An ordered vector space R is homogeneous in the sense that for each two points $p, q \in R$ there is a definable homeomorphism from R onto itself that sends p

to q, namely $x \mapsto x+(q-p): R \to R$. This homogeneity does not necessarily extend from points to intervals, as the following example shows.

EXAMPLE. Equip the cartesian product $\mathbb{R} \times \mathbb{Q}$ of additive groups with the lexicographic ordering:

$$(r_1, q_1) < (r_2, q_2) \text{ iff } r_1 < r_2, \text{ or } r_1 = r_2 \text{ and } q_1 < q_2.$$

This makes $\mathbb{R} \times \mathbb{Q}$ into a divisible ordered abelian group in which there are both countable intervals and uncountable intervals. In particular, not every two intervals of $\mathbb{R} \times \mathbb{Q}$ are homeomorphic, let alone definably homeomorphic. This example is used later to show that certain results on o-minimal structures expanding ordered fields do not hold in general for o-minimal structures expanding ordered abelian groups.

(7.12) DIGRESSION: THE DIVISION RING OF DEFINABLE ADDITIVE MAPS. We have just seen that ordered vector spaces over ordered fields, and, more generally, over ordered division rings, are o-minimal. Here we show, using a result from Chapter 3, that any o-minimal structure that expands an ordered abelian group is in fact an ordered vector space over an ordered division ring that is canonically associated to the structure.

Let $(R, <, \mathcal{S})$ be an o-minimal structure which expands an ordered abelian group $(R, <, 0, -, +)$.

Let $\lambda : R \to R$ be a definable additive map. (Additive: $\lambda(x + y) = \lambda(x) + \lambda(y)$.) Since λ is definable, the monotonicity theorem from Chapter 3 implies that λ is continuous at some point of R. Then additivity of λ implies that λ is continuous on all of R. Since R has no definable subgroups besides $\{0\}$ and itself it follows by looking at the kernel and the image of λ that either $\lambda = 0$ or λ is a bijection. Suppose λ is a bijection. Then $(0, \infty)$ is the disjoint union of the two definable open sets $\{x > 0 : \lambda(x) > 0\}$ and $\{x > 0 : \lambda(x) < 0\}$, so one of these two sets is all of $(0, \infty)$. In other words, either $\lambda(x) > 0$ for all $x > 0$, or $\lambda(x) < 0$ for all $x > 0$. Write $\lambda > 0$ in the first case and $\lambda < 0$ in the second case. The set of definable additive maps on R is a ring under pointwise addition and composition with identity 1_R, and we have just defined an ordering on this ring. We have shown

PROPOSITION. *The ring of definable additive maps on R is an ordered division ring.*

Hence R is naturally an ordered vector space over this division ring, and the sets definable in this ordered vector space using constants are clearly also definable in $(R, <, \mathcal{S})$. There is more to say in this direction, see [21, §4].

EXERCISE.

Let R be a nontrivial ordered vector space over an ordered division ring F, and consider R as an L_F-structure. Show that the maps $R \to R$ that are additive and definable using constants are exactly the scalar multiplications by elements of F. (Hence these maps are actually definable without using constants.)

Notes and comments

The results in Section 4 on o-minimal ordered groups and rings are from Pillay and Steinhorn [49]. (There are other notions of "ordered ring" in use that allow nonzero nilpotents, cf. Fuchs [25].) The content of Section 6 is also known among model-theorists as "elimination of quantifiers" for dense linearly ordered sets, cf. Chang & Keisler [9]. The key argument in the proof of Proposition 7.4 is a form of "Fourier-Motzkin elimination", and thus goes back a long time. See also Sontag [55].

The results in sections 4, 6 and 7 already hint towards a division of all o-minimal structures into three fundamentally different kinds:

(1) Those in which very few sets are definable, like the dense linearly ordered sets without endpoints.
(2) Non-trivial ordered vector spaces over ordered division rings.
(3) O-minimal expansions of real closed fields.

A precise result along these lines, the Trichotomy Theorem for o-minimal structures, has indeed been established by Peterzil and Starchenko [46]. This theorem can be seen as a manifestation of the so-called Zil'ber Principle of geometric model-theory, which clarifies the intrinsic nature of such classifications. These matters fall unfortunately outside the scope of this book.

CHAPTER 2

SEMIALGEBRAIC SETS

Introduction

In this chapter we prove that the sets definable with parameters in the field $\mathbb{R}$ are just the semialgebraic sets. This characterization of the definable sets is essentially the Tarski-Seidenberg theorem. Here we give Łojasiewicz's proof, which gives rather precise information on semialgebraic sets. The ideas involved will also be useful when we deal with semianalytic sets and their generalizations in the next volume. Łojasiewicz's proof was an eye-opener to me, and motivated the notions of o-minimal structure and cell decomposition. Section 1 contains some lemmas that are also useful elsewhere. In Section 2 we prove the main theorem (2.7), and Section 3 characterizes closed semialgebraic sets.

§1. Thom's lemma and continuity of roots

We start with a simple but useful bound on the zeros of a polynomial with complex coefficients:

(1.1) LEMMA. *Let $\alpha\in\mathbb{C}$ be a zero of the monic polynomial*

$$a_0 + a_1T + \cdots + a_{d-1}T^{d-1} + T^d \in \mathbb{C}[T],\ d \geq 1.$$

Then $|\alpha| \leq 1 + \max\{|a_i| : i = 0,\ldots,d-1\}$.

PROOF. Put $M := \max\{|a_i| : i = 0,\ldots,d-1\}$ and suppose $|\alpha| > 1 + M$. Then $|a_0+a_1\alpha+\cdots+a^{d-1}\alpha^{d-1}| \leq M\cdot(1+|\alpha|+\cdots+|\alpha|^{d-1}) = M\cdot(|\alpha|^d-1)/(|\alpha|-1) < |\alpha|^d$, contradicting $0 = |f(\alpha)| = |a_0 + a_1\alpha + \cdots + a_{d-1}\alpha^{d-1} + \alpha^d|$. □

(1.2) THOM'S LEMMA. *Let $f_1,\ldots,f_k \in \mathbb{R}[T]$ be nonzero polynomials such that if $f_i' \neq 0$, then $f_i'\in\{f_1,\ldots,f_k\}$. Let $\epsilon\colon\{1,\ldots,k\}\to\{-1,0,1\}$, and put*

$$A_\epsilon := \{t\in\mathbb{R} : \operatorname{sign}(f_i(t)) = \epsilon(i),\ i = 1,\ldots,k\},\ \textit{a subset of}\ \mathbb{R}.$$

Then A_ϵ is empty, a point, or an interval. If $A_\epsilon \neq \emptyset$, then its closure is given by

$$\operatorname{cl}(A_\epsilon) = \{t\in\mathbb{R} : \operatorname{sign}(f_i(t)) \in \{\epsilon(i),0\},\ i = 1,\ldots,k\}.$$

If $A_\epsilon = \emptyset$, then $\{t \in \mathbb{R} : \operatorname{sign}(f_i(t)) \in \{\epsilon(i), 0\},\ i = 1, \ldots, k\}$ is empty or a point.

REMARK. We may call ϵ a **sign condition** for $f_1, \ldots, f_k$. The 3^k possible sign conditions ϵ determine 3^k disjoint sets A_ϵ, which together cover the real line $\mathbb{R}$. Of course for many ϵ's the corresponding A_ϵ may be empty. The second statement of the lemma says that for nonempty A_ϵ its closure can be obtained by relaxing all strict inequalities to weak inequalities.

PROOF OF THOM'S LEMMA. By induction on k. The lemma holds trivially for $k = 0$. Let $f_1, \ldots, f_k, f_{k+1} \in \mathbb{R}[T] - \{0\}$ be polynomials such that if $f_i' \neq 0$, then $f_i' \in \{f_1, \ldots, f_{k+1}\}$. We may as well assume that $\deg(f_{k+1}) = \max\{\deg(f_i) : 1 \leq i \leq k+1\}$. Let $\epsilon' : \{1, \ldots, k+1\} \to \{-1, 0, 1\}$, and let ϵ be the restriction of ϵ' to $\{1, \ldots, k\}$. By the inductive hypothesis A_ϵ is empty, a point, or an interval. If A_ϵ is empty or a point, so is $A_{\epsilon'} := A_\epsilon \cap \{t \in \mathbb{R} : \operatorname{sign}(f_{k+1}(t)) = \epsilon'(k+1)\}$, and the other properties to be checked in this case follow easily from the inductive hypothesis on A_ϵ. Suppose A_ϵ is an interval. Since f_{k+1}' has a constant sign on A_ϵ, the function f_{k+1} is either strictly monotone on A_ϵ, or constant. In both cases it is routine to check that $A_{\epsilon'} = A_\epsilon \cap \{t \in \mathbb{R} : \operatorname{sign}(f_{k+1}(t)) = \epsilon'(k+1)\}$ has the required properties. □

(1.3) LEMMA (CONTINUITY OF ROOTS). *Let $f(T) = a_0 + a_1T + \cdots + a_dT^d \in \mathbb{C}[T]$ be a polynomial that has no zero on the boundary circle $|z - c| = r$ of a given open disc $|z - c| < r$ in the complex plane ($c \in \mathbb{C}$, $r > 0$). Then there is $\epsilon > 0$ such that if $|a_i - b_i| \leq \epsilon$ for $i = 0, \ldots, d$, then $g(T) := b_0 + b_1T + \cdots + b_dT^d \in \mathbb{C}[T]$ also has no zero on the circle, and f and g have the same number of zeros in the disc.* (Here and in the proof below we count zeros *with* their multiplicity.)

PROOF. From complex analysis we know that the number of zeros of f in the disc $|z - c| < r$ equals

$$\frac{1}{2\pi \mathrm{i}} \int_C \frac{f'(z)}{f(z)} \mathrm{d}z,$$

where the circle C is parametrized counterclockwise as $c + re^{2\pi \mathrm{i}\theta}$, $0 \leq \theta \leq 1$. By taking ϵ sufficiently small we can clearly guarantee that g has no zeros on C as well, and then the number of zeros of g in the disc is given by the corresponding integral of $(1/2\pi\mathrm{i})(g'(z)/g(z))\mathrm{d}z$ over C. Moreover, for small enough ϵ the difference $|(f'/f) - (g'/g)|$ will be so small everywhere on C that the values of the two integrals over C must differ by less than 1 in absolute value. Since these values are integers, they must be equal. □

(1.4) EXERCISES.

1. Let $f = f(X_1, \ldots, X_m) \in F[X_1, \ldots, X_m]$ be a polynomial with coefficients in the field F, and let $d_1, \ldots, d_m \in \mathbb{N}$ be such that $\deg_{X_i} f \leq d_i$ for $i = 1, \ldots, m$. Show that if f vanishes identically on a cartesian product $A_1 \times \cdots \times A_m$ with $|A_1| > d_1, \ldots, |A_m| > d_m$ (all $A_i \subseteq F$), then $f = 0$.

2. Let F be an ordered field and $f \in F[X_1,\dots,X_m]$, $f \neq 0$. Show that the zero set $Z(f) := \{a \in F^m : f(a) = 0\}$ is a closed subset of F^m with empty interior.

§2. Semialgebraic cell decomposition

(2.1) We now come to the setting for Lojasiewicz's theorem:

X is a nonempty topological space, $\mathbf{E}$ a ring of continuous real-valued functions $f: X \to \mathbb{R}$, the ring operations being pointwise addition and multiplication, with multipicative identity the function on X that takes the constant value 1.

Call a set $A \subseteq X$ an **E**-set if A is a finite union of sets of the form

$$\{x \in X : f(x) = 0,\ g_1(x) > 0, \dots, g_k(x) > 0\}, \text{ with } f, g_1, \dots, g_k \in \mathbf{E}.$$

Note that the **E**-sets form a boolean algebra of subsets of X.

(2.2) EXAMPLE. If $X = \mathbb{R}^m$, and $\mathbf{E} = \mathbb{R}[x_1,\dots,x_m]$ is the ring of real polynomial functions on X, then the **E**-sets are exactly the semialgebraic subsets of $\mathbb{R}^m$.

(2.3) Note that the pair $(X \times \mathbb{R}, \mathbf{E}[T])$ also satisfies the conditions imposed on $(X, \mathbf{E})$, where $X \times \mathbb{R}$ is given the product topology, and each polynomial

$$f(T) = f_0 + f_1 T + \cdots + f_d T^d \in \mathbf{E}[T]$$

is interpreted as the (continuous) function $(x,t) \mapsto f_d(x)t^d + \cdots + f_0(x) : X \times \mathbb{R} \to \mathbb{R}$.

NOTATION. For $f(T)$ as above and $x \in X$, we write $f(x,T)$ for the polynomial

$$f(x,T) = f_0(x) + f_1(x)T + \cdots + f_d(x)T^d \in \mathbb{R}[T].$$

(2.4) LEMMA. *Suppose X is connected. Let $f = f_0 + f_1 T + \cdots + f_d T^d \in \mathbf{E}[T]$, and suppose $e \leq d$ is such that the polynomial $f(x,T) \in \mathbb{R}[T]$ has exactly e distinct complex zeros, for each $x \in X$. Then the number of distinct real zeros of $f(x,T)$ is also constant as x ranges over X. Writing $\zeta_1(x) < \cdots < \zeta_r(x)$ for these real zeros, the functions $\zeta_j : X \to \mathbb{R}$ are continuous.*

REMARK. Zeros are counted here *without* multiplicity.

PROOF. Let $x_0 \in X$ and let $z_1, \dots, z_e$ be the distinct complex zeros of $f(x_0, T)$. Take closed balls B_i centered at z_i in the complex plane $\mathbb{C}$, such that $B_i \cap B_j = \emptyset$ for $i \neq j$, and $B_i \cap \mathbb{R} = \emptyset$ if $z_i \notin \mathbb{R}$. By "continuity of roots" there is a neighborhood U of x_0 in X such that for each $x \in U$ the ball B_i contains at least one zero $\zeta_i(x)$ of $f(x,T)$. Note that then $\zeta_i(x)$ is in fact the only zero of $f(x,T)$ in B_i. The graph of ζ_i on U is $\{(x,t)\in U\times B_i:\ f(x,t)=0\}$, hence this graph is closed in $U\times B_i$; in combination with the compactness of B_i this implies that ζ_i is continuous on U.

Since the coefficients of $f(x,T)$ are real, the set $\{\zeta_1(x),\dots,\zeta_e(x)\}$ is closed under complex conjugation. Hence, if $\zeta_i(x_0)=z_i\in\mathbb{R}$, then $\zeta_i(x)\in\mathbb{R}$ for all $x\in U$. This shows that the number of real zeros is locally constant, and thus constant, since X is connected. The real $\zeta_i(x)$'s must also keep their order on the real line as x runs through U. □

(2.5) The lemma below is a special case of Chevalley's constructibility theorem, also known as quantifier elimination for algebraically closed fields, to be proved in the next volume. We give here Łojasiewicz's direct proof of this special case.

(2.6) LEMMA. *Let $A=(A_0,\dots,A_d)$ be a tuple of distinct variables and let*

$$f(A,T)=A_0+A_1T+\cdots+A_dT^d\in\mathbf{Z}[A,T]$$

be the general polynomial of degree d. Let $e\in\{0,\dots,d\}\cup\{\infty\}$. Then the set

$$\{a=(a_0,\dots,a_d)\in\mathbf{C}^{d+1}:\ f(a,T)\text{ has exactly } e \text{ distinct complex zeros}\}$$

is a finite union of sets of the form

$$\{a\in\mathbf{C}^{d+1}:\ p_1(a)=\cdots=p_k(a)=0, q(a)\neq 0\},\ \textit{where } p_i(A),\ q(A)\in\mathbf{Z}[A].$$

Note that for $e=0$ the set is $\{(a_0,0,\dots,0):\ a_0\neq 0\}$; for $e=\infty$ the set is just the one-point set $\{(0,0,\dots,0)\}$. This lemma implies in particular that for $f(T)=f_0+f_1T+\cdots+f_dT^d\in\mathbf{E}[T]$ the set $\{x\in X:\ f(x,T)$ has exactly e complex zeros$\}$ is an $\mathbf{E}$-set.

PROOF OF (2.6). Let $d>0$, and $(a_0,\dots,a_d)\in\mathbf{C}^{d+1}$ with $a_d\neq 0$. Put $g:=$ degree of $\gcd\big(f(a,T),(\partial f/\partial T)(a,T)\big)$ in $\mathbf{C}[T]$; then the lcm of these two polynomials has degree $2d-g-1$, and the number of distinct complex zeros of $f(a,T)$ is $d-g$. Let $0<k<d$. Hence the condition

$$(*)\qquad \begin{cases} f(a,T)\cdot q(x,T)=\big(\partial f/\partial T\big)(a,T)\cdot r(x,T)\text{ for some nonzero}\\ x=(x_0,\dots,x_{2k})\in\mathbf{C}^{2k+1},\text{ where}\\ q(x,T)=x_0+x_1T+\cdots+x_{k-1}T^{k-1}\text{ and}\\ r(x,T)=x_k+x_{k+1}T+\cdots+x_{2k}T^k\end{cases}$$

is equivalent to $2d-g-1\le d+k-1$, that is, to $d-g\le k$, that is, to the condition that $f(a,T)$ has at most k distinct zeros. Clearly we have

$$\begin{aligned}&f(a,T)\cdot q(x,T)-\big(\partial f/\partial T\big)(a,T)\cdot r(x,T)\\&=\beta_0(a,x)+\beta_1(a,x)T+\cdots+\beta_{d+k-1}(a,x)T^{d+k-1}\end{aligned}$$

for certain bilinear functions $\beta_0, \ldots, \beta_{d+k-1} : \mathbf{C}^{d+1} \times \mathbf{C}^{2k+1} \to \mathbf{C}$.

Hence $(*)$ is equivalent to the condition that $\beta_i(a, x) = 0$ for all i, and some nonzero $x \in \mathbf{C}^{2k+1}$, that is, to the condition that the linear map

$$x \mapsto \big(\beta_0(a,x), \ldots, \beta_{d+k-1}(a,x)\big) : \mathbf{C}^{2k+1} \to \mathbf{C}^{d+k}$$

has nontrivial kernel. This last condition is equivalent to the vanishing of all $(2k+1) \times (2k+1)$ minors of the matrix of this linear map. This exhibits the set

$$\big\{a \in \mathbf{C}^{d+1} \ : \ a_d \neq 0 \text{ and } f(a,T) \text{ has at most } k \text{ distinct complex zeros}\big\}$$

as the intersection of the set $\big\{a \in \mathbf{C}^{d+1} \ : \ a_d \neq 0\big\}$ with the zero set of certain polynomials in $\mathbf{Z}[A]$. The desired result now follows easily. □

(2.7) Theorem. *Let $f_1(T), \ldots, f_M(T) \in \mathbf{E}[T]$. Then the list $f_1, \ldots, f_M$ can be augmented to a list $f_1, \ldots, f_N$ in $\mathbf{E}[T]$ $(M \leq N)$, and X can be partitioned into finitely many $\mathbf{E}$-sets X_i $(1 \leq i \leq k)$ such that for each connected component C of each X_i there are continuous real-valued functions $\xi_{C,1} < \cdots < \xi_{C,m(C)}$ on C with the following two properties:*

1. *each function f_n $(1 \leq n \leq N)$ has constant sign $(-1, 0,$ or $1)$ on each of the graphs $\Gamma(\xi_{C,j})$ $(1 \leq j \leq \mu(C))$ and on each of the sets $(\xi_{C,j}, \xi_{C,j+1})$ $(0 \leq j \leq \mu(C))$, where $\xi_{C,0} := -\infty$ and $\xi_{C,m(C)+1} := +\infty$ are constant functions on C by convention;*
2. *each of the sets $\Gamma(\xi_{C,j})$ and $(\xi_{C,j}, \xi_{C,j+1})$ from (1) is of the form*

$$\big\{(x,t) \in C \times \mathbf{R} : \ \operatorname{sign}\big(f_n(x,t)\big) = \epsilon(n) \text{ for } n = 1, \ldots, N\big\}$$

for a suitable sign condition $\epsilon : \{1, \ldots, N\} \to \{-1, 0, 1\}$.

Proof. Take $d \in \mathbf{N}$ such that each of the polynomials $f_m(T)$ is of the form

$$f_m(T) = f_{m0} + f_{m1}T + \cdots + f_{md}T^d, \text{ with } f_{mr} \in \mathbf{E} \text{ for } 0 \leq r \leq d.$$

Let Δ range over the subsets of $\{1, \ldots, M\} \times \{0, \ldots, d\}$, and put

$$f_\Delta(T) := \prod_{(m,r) \in \Delta} (\partial^r f_m / \partial T^r) \in \mathbf{E}[T], \text{ so } \deg_T(f_\Delta) \leq Md^2.$$

We let e range over $\{0, \ldots, Md^2, \infty\}$, and put $A_{\Delta e} := \big\{x \in X \ : \ f_\Delta(x,T) \text{ has exactly } e \text{ complex zeros}\big\}$. From the remark following lemma (2.6) we obtain that $A_{\Delta e}$ is an $\mathbf{E}$-set. Clearly, for a given Δ the sets $A_{\Delta e}$ are pairwise disjoint as e varies, and cover X. Because the $\mathbf{E}$-sets form a boolean algebra we can find a partition $X = X_1 \cup \cdots \cup X_k$ of X into $\mathbf{E}$-sets X_i such that each set $A_{\Delta e}$ is a union of X_i's. Augment $f_1, \ldots, f_M$ to $f_1, \ldots, f_N$ such that

$$\{f_1, \ldots, f_N\} = \big\{\partial^r f_m / \partial T^r \ : \ 1 \leq m \leq M, \ 0 \leq r \leq d\big\}.$$

We shall prove that this sequence $f_1, \ldots, f_N$ and these X_i's satisfy the conclusion of our theorem.

For each connected component C of an X_i, let $\Delta(C)$ be the set

$$\left\{(m,r) \in \{1,\ldots,M\} \times \{0,\ldots,d\} : \frac{\partial^r f_m}{\partial T^r} \text{ does not vanish identically on } C \times \mathbb{R}\right\}.$$

Fix such a component C. Given any Δ, there exists e such that $C \subseteq A_{\Delta e}$, in particular C is contained in $A_{\Delta(C)e}$ for some e. Then $f_{\Delta(C)}(x,T)$ has exactly e complex zeros, for each $x \in C$. We claim e is finite. To see this, consider a factor $g(T) = \partial^r f_m/\partial T^r$ of $f_{\Delta(C)}$, $(m,r) \in \Delta(C)$. The set C is contained in $A_{\{(m,r)\}e'}$ for some $e' \in \{0,\ldots,d,\infty\}$, so the number e' of complex zeros of $g(x,T)$ is independent of $x \in C$. Because $(m,r) \in \Delta(C)$, this number e' is finite, and hence e is finite. Lemma (2.4) now implies there are continuous real-valued functions $\xi_{C,1} < \cdots < \xi_{C,\mu(C)}$ on C such that

$$\{(x,t) \in C \times \mathbb{R} : f_{\Delta(C)}(x,t) = 0\} = \Gamma(\xi_{C,1}) \cup \cdots \cup \Gamma(\xi_{C,\mu(C)}).$$

We shall prove that these functions $\xi_{C,j}$ satisfy the conclusions of the theorem.

CLAIM 1. Each function $\partial^r f_m/\partial T^r$ $(1 \le m \le M,\ 0 \le r \le d)$ has constant sign on each set $\Gamma(\xi_{C,j})$ $(1 \le j \le \mu(C))$ and $(\xi_{C,j}, \xi_{C,j+1})$ $(0 \le j \le \mu(C))$, where $\xi_{C,0} = -\infty$ and $\xi_{C,\mu(C)+1} = +\infty$.

Note that this claim gives us conclusion (1) of the theorem, since each f_n is one of the $\partial^r f_m/\partial T^r$'s. To prove the claim, note first that $\partial^r f_m/\partial T^r$ vanishes identically on the indicated sets if $(m,r) \notin \Delta(C)$. Let $(m,r) \in \Delta(C)$. Then, with $g(T) := \partial^r f_m/\partial T^r$, the number of complex zeros of $g(x,T)$ is independent of $x \in C$, as we already saw. Hence by lemma (2.4) the zero set of g on $C \times \mathbb{R}$ is a finite disjoint union of graphs of continuous real-valued functions on C. Since $f_{\Delta(C)}$ contains g as a factor, it is clear that these functions must be among the $\xi_{C,j}$'s. (Use that C is connected.) This proves claim 1.

Now let B be one of the sets in claim 1. Put $\epsilon(m,r) := \operatorname{sign}(\partial^r f_m/\partial T^r)$ on B, and put

$$B' := \left\{(x,t) \in C \times \mathbb{R} : \operatorname{sign}\left(\frac{\partial^r f_m}{\partial T^r}\right)(x,t) = \epsilon(m,r) \text{ for } 1 \le m \le M,\ 0 \le r \le d\right\}.$$

Clearly $B \subseteq B'$.

CLAIM 2. $B = B'$.

Suppose not. Take $(x,t') \in B' - B$ and $(x,t) \in B$. Say $t < t'$. Thom's lemma implies that $\{s \in \mathbb{R} : (x,s) \in B'\}$ is connected, so $\{x\} \times [t,t'] \subseteq B'$. Note that $f_{\Delta(C)}$ must change sign on $\{x\} \times [t,t']$ since $(x,t) \in B$ and $(x,t') \notin B$. But $f_{\Delta(C)}$ is a product of partials $\partial^r f_m / \partial T^r$, so $f_{\Delta(C)}$ cannot change sign on B', contradiction. This proves claim 2, and thus conclusion (2) of the theorem. □

(2.8) Let us say that the pair $(X, \mathbf{E})$ has the **Łojasiewicz property** if each $\mathbf{E}$-set has only finitely many connected components, and each component is also an $\mathbf{E}$-set.

(2.9) COROLLARY. *If $(X, \mathbf{E})$ has the Łojasiewicz property, then $(X \times \mathbb{R}, \mathbf{E}[T])$ also has the Łojasiewicz property, and the image of any $\mathbf{E}[T]$-set $S \subseteq X \times \mathbb{R}$ under the projection map $X \times \mathbb{R} \to X$ is an $\mathbf{E}$-set.*

PROOF. Write an $\mathbf{E}[T]$-set S as a finite union of sets of the form

$$\{(x,t) \in X \times \mathbb{R} : f(x,t) = 0, \ g_1(x,t) > 0, \ldots, g_k(x,t) > 0\}$$

with $f, g_1, \ldots, g_k \in \mathbf{E}[T]$. Let $f_1, \ldots, f_M$ be a list of all the polynomials in $\mathbf{E}[T]$ that are involved in such a definition of S. Now apply the theorem to this list. □

(2.10) The $\mathbb{R}[T_1, \ldots, T_m]$-sets in the space $\mathbb{R}^m$ are exactly the semialgebraic subsets of $\mathbb{R}^m$. For $m = 0$ the space $\mathbb{R}^m$ has only one point, so the pair $(\mathbb{R}^m, \mathbb{R}[T_1, \ldots, T_m])$ obviously has the Łojasiewicz property for $m = 0$; hence by induction on m, using (2.9), it follows that $(\mathbb{R}^m, \mathbb{R}[T_1, \ldots, T_m])$ has the Łojasiewicz property for each m, and that the image of a semialgebraic subset of $\mathbb{R}^{m+1}$ under the projection map $\mathbb{R}^{m+1} \to \mathbb{R}^m$ is a semialgebraic subset of $\mathbb{R}^m$. (Tarski-Seidenberg property)

We can now draw the following conclusion about the model-theoretic structure $(\mathbb{R}, <, 0, 1, +, -, \cdot)$, which we will call "the ordered field of real numbers".

(2.11) COROLLARY. *The sets $S \subseteq \mathbb{R}^m$ $(m = 0, 1, 2, \ldots)$ that are definable using constants in the ordered field of real numbers are exactly the semialgebraic sets. The ordered field of real numbers is o-minimal.*

PROOF. By (2.10) the semialgebraic sets form a structure on $(\mathbb{R}, <)$. Clearly the primitives of the ordered field of reals are semialgebraic. By Chapter 1, (5.9), exercise 3, the semialgebraic sets are definable using constants in the ordered field of real numbers. Finally, the semialgebraic subsets of $\mathbb{R}$ are clearly the finite unions of intervals and points. The desired result follows. □

This classical example should be kept in mind throughout this book. In the next volume we extend this result by showing that all real closed fields are o-minimal, in a uniform way, and indicate how this fact throws light on the classical real case as well. There we also treat the remarkable recent theorem by Wilkie that the ordered *exponential* field of reals is o-minimal, and its aftermath.

§3. Thom's lemma with parameters

This last section is mainly a preparation for our later work on semianalytic sets in a later volume. We return to the set-up where we have a pair $(X, \mathbf{E})$ as before: X a nonempty topological space, $\mathbf{E}$ a ring of real-valued continuous functions on X.

(3.1) DEFINITION. Let $f_1(T), \dots, f_M(T) \in \mathbf{E}[T]$, and let C be a connected subset of X. A **decomposition of $f_1, \dots, f_M$ above C** consists of continuous real-valued functions $\xi_1 < \dots < \xi_r$ on C such that

(1) each function f_m has constant sign on each set $\Gamma(\xi_i)$ $(1 \le i \le r)$ and on each set (ξ_i, ξ_{i+1}) $(0 \le i \le r$, where $\xi_0 := -\infty$, $\xi_{r+1} := +\infty)$;

(2) each set $\Gamma(\xi_i)$ is contained in the zero set of some function f_m.

Note: if (1) holds, then (2) can be satisfied by removing the ξ_i's not satisfying (2). Theorem (2.7) implies that if $(X, \mathbf{E})$ has the Łojasiewicz property, then for every sequence $f_1, \dots, f_M$ in $\mathbf{E}[T]$ there is a partition of X into finitely many connected $\mathbf{E}$-sets, above each of which $f_1, \dots, f_M$ has a decomposition.

Let us say that a polynomial $f(T) \in \mathbf{E}[T]$ **has zero-free leading coefficient** if $f(T) = c_d T^d + c_{d-1}T^{d-1} + \dots + c_0$ with $c_i \in \mathbf{E}$, such that c_d has no zero on X.

By (1.1) the real zeros of a polynomial $a_d T^d + a_{d-1}T^{d-1} + \dots + a_0 \in \mathbb{R}[T]$ with $a_d \neq 0$ are in absolute value $\le 1 + \max\{|a_i/a_d| : 0 \le i \le d-1\}$.

Hence if $f(T) \in \mathbf{E}[T]$ has zero-free leading coefficient and a point $y \in X$ is given, then the real zeros of $f(x, T)$ are bounded in absolute value by some constant, for all x in a suitably small neighborhood of y.

This fact is used in the proof of the next lemma, which constitutes what one might call "Thom's lemma with parameters".

(3.2) LEMMA. *Suppose the polynomials $f_1, \dots, f_N \in \mathbf{E}[T]$ have zero-free leading coefficients, and each nonzero partial $\partial f_n/\partial T$ also belongs to $\{f_1, \dots, f_N\}$. Let $\xi_1 < \dots < \xi_r$ be a decomposition of $f_1, \dots, f_N$ above a connected set $C \subseteq X$. Fix $i \in \{1, \dots, r\}$, let $\xi = \xi_i$, put $\epsilon(n) = \operatorname{sign}(f_n|\Gamma(\xi))$ for $1 \le n \le N$. Then*

(a) $\Gamma(\xi) = \{(x,t) \in C \times \mathbb{R} : \operatorname{sign}(f_n(x,t)) = \epsilon(n) \text{ for } n = 1, \dots, N\}$,

(b) $\xi : C \to \mathbb{R}$ *extends uniquely to a continuous function* $\eta : \operatorname{cl}(C) \to \mathbb{R}$, *and* $\operatorname{cl}(\Gamma(\xi)) = \Gamma(\eta) = \{(y,t) \in \operatorname{cl}(C) \times \mathbb{R} : \operatorname{sign}(f_n(y,t)) \in \{\epsilon(n), 0\} \text{ for } n = 1, \dots, N\}$.

PROOF. For fixed $x \in C$ the set

$$A_{x,\epsilon} := \{t \in \mathbb{R} : \operatorname{sign}(f_n(x,t)) = \epsilon(n) \text{ for } n = 1, \dots, N\}$$

contains $\{\xi(x)\}$ and is finite since some f_m vanishes on $\Gamma(\xi)$ and $f_m(x,T) \neq 0$. The polynomials $f_1(x,T), \dots, f_N(x,T)$ satisfy the hypothesis of Thom's lemma, so $A_{x,\epsilon}$ is connected. Hence $A_{x,\epsilon} = \{\xi(x)\}$. This proves (a).

For (b), take m such that $f_m(x, \xi(x)) = 0$ for all $x \in C$. Let $y \in \mathrm{cl}(C)$. Since $f_m(T)$ has zero-free leading coefficient, there is a neighborhood U of y such that ξ is bounded on $U \cap C$. Hence the limits

$$l := \liminf_{x \to y} \xi(x)$$
$$\text{and } L := \limsup_{x \to y} \xi(x)$$

are real numbers (not $\pm\infty$). Now $\mathrm{sign}(f_n(x, \xi(x))) = \epsilon(n)$ for all n, hence $\mathrm{sign}(f_n(y, l)) \in \{\epsilon(n), 0\}$, and $\mathrm{sign}(f_n(y, L)) \in \{\epsilon(n), 0\}$ for all n. By Thom's lemma the set $\{t \in \mathbb{R} : \mathrm{sign}(f_n(y,t)) \in \{\epsilon(n), 0\} \text{ for } n = 1, \ldots, N\}$ is connected, so this set contains $[l, L]$. With m as above we have $f_m(y,t) = 0$ for all $t \in [l, L]$, and because $f_m(y, T) \neq 0$, this implies $l = L$. Since y was an arbitrary point of $\mathrm{cl}(C)$ we have shown that $\eta(y) := \lim_{x \to y} \xi(x)$ defines a function $\eta : \mathrm{cl}(C) \to \mathbb{R}$. Continuity of ξ implies continuity of η, so $\Gamma(\eta)$ is closed, hence $\mathrm{cl}(\Gamma(\xi)) = \Gamma(\eta)$.

The argument with Thom's lemma that we just gave also shows

$$\Gamma(\eta) = \{(y,t) \in \mathrm{cl}(C) \times \mathbb{R} : \mathrm{sign}(f_n(y,t)) \in \{\epsilon(n), 0\} \text{ for } n = 1, \ldots, N\}.$$

This finishes the proof of (b). □

(3.3) Lemma. *With the same assumptions as in the previous lemma, set* $\xi_0 := -\infty$, $\xi_{r+1} := +\infty$. *Fix an* $i \in \{0, \ldots, r\}$ *and put* $\epsilon(n) = \mathrm{sign}(f_n|(\xi_i, \xi_{i+1}))$. *Then we have*

(a) $(\xi_i, \xi_{i+1}) = \{(x,t) \in C \times \mathbb{R} : \mathrm{sign}(f_n(x,t)) = \epsilon(n) \text{ for } n = 1, \ldots, N\}$,
(b) $\mathrm{cl}(\xi_i, \xi_{i+1}) = \{(y,t) \in \mathrm{cl}(C) \times \mathbb{R} : \mathrm{sign}(f_n(y,t)) \in \{\epsilon(n), 0\} \text{ for } n = 1, \ldots, N\}$.

We leave this as an exercise.

(3.4) Further terminology. Let F be a finite set of continuous real-valued functions on a topological space X. We say that a set $S \subseteq X$ is **described by** F if S is a finite union of sets of the form

$$\epsilon(F) := \{x \in X : \mathrm{sign}(f(x)) = \epsilon(f) \text{ for all } f \in F\}, \text{ with } \epsilon : F \to \{-1, 0, 1\}.$$

Since F is a finite collection of functions, the sets described by F form a finite boolean algebra of subsets of X.

(3.5) Noether normalization. Let $f(X_1, \ldots, X_n)$ be a nonzero polynomial of degree d over a field K of characteristic 0, say $f = \sum_{|i|=d} c_i X^i$ + terms of lower degree, $n > 0$. Here $i = (i_1, \ldots, i_n) \in \mathbb{N}^n$, $|i| = i_1 + \cdots + i_n$ and $X^i = X_1^{i_1} \cdots X_n^{i_n}$, $c_i \in K$. Let $(\lambda_1, \ldots, \lambda_{n-1}) \in K^{n-1}$. We are going to make the (invertible) substitution

$$\begin{aligned} X_1 &\to X_1 + \lambda_1 X_n, \\ &\vdots \\ X_{n-1} &\to X_{n-1} + \lambda_{n-1} X_n, \\ X_n &\to X_n. \end{aligned}$$

Note that, with $\lambda^i = \lambda_1^{i_1} \cdots \lambda_{n-1}^{i_{n-1}}$, this substitution transforms f into

$$f(X_1 + \lambda_1 X_n, \ldots, X_{n-1} + \lambda_{n-1} X_n, X_n)$$
$$= \left(\sum_{|i|=d} c_i \lambda^i \right) X_n^d + \text{ terms of degree } < d \text{ in } X_n.$$

Since the polynomial $\sum_{|i|=d} c_i Y_1^{i_1} \cdots Y_{n-1}^{i_{n-1}} \in K[Y_1, \ldots, Y_{n-1}]$ is nonzero, it follows that for "almost all" $\lambda = (\lambda_1, \ldots, \lambda_{n-1}) \in K^{n-1}$, namely for all those λ that are not zeros of this polynomial, the transformed polynomial

$$f(X_1 + \lambda_1 X_n, \ldots, X_{n-1} + \lambda_{n-1} X_n, X_n)$$

is of (total) degree d, and has a term of the form cX_n^d with $c \in K - \{0\}$.

(3.6) COROLLARY. *A closed semialgebraic subset of* $\mathbb{R}^n$ *is a finite union of sets of the form* $\{x \in \mathbb{R}^n : f_1(x) \geq 0, \ldots, f_k(x) \geq 0\}$ *with* $f_1, \ldots, f_k \in \mathbb{R}[X_1, \ldots, X_n]$.

PROOF. By induction on n. The case $n = 0$ is trivial; assume the desired result holds for a certain n, let $S \subseteq \mathbb{R}^{n+1}$ be a closed semialgebraic set, and let F be a finite set of nonzero polynomials in $\mathbb{R}[X_1, \ldots, X_n, T]$ describing S. By a linear transformation of variables as in (3.5) we may as well assume that all $f \in F$ are monic in T up to a constant factor, and hence, by enlarging F, we may also assume that each nonzero partial $\partial^r f / \partial T^r$ of each $f \in F$ belongs to F. By theorem (2.7) and by (2.10) there is a partition $\mathcal{P}$ of $\mathbb{R}^n$ into finitely many connected semialgebraic sets C such that F has a decomposition over each $C \in \mathcal{P}$. Now apply the inductive assumption to the closed semialgebraic sets $\text{cl}(C)$, and apply lemmas (3.2) and (3.3) to get the desired result. □

(3.7) EXERCISES.

1. Let $Q(X, Y) \in \mathbb{R}[X, Y]$ be a nonzero polynomial in two variables. Show that there are $d \in \mathbb{N}$ and $M > 0$ such that if $(x, y) \in \mathbb{R}^2$, $x > M$ and $Q(x, y) = 0$, then $|y| \leq x^d$.

In the next two problems we say that a map $g : A \to \mathbb{R}^n$ with $A \subseteq \mathbb{R}^m$ is **semialgebraic** if its graph $\Gamma(g) \subseteq \mathbb{R}^{m+n}$ is semialgebraic.

2. Show that if $g : \mathbb{R} \to \mathbb{R}$ is semialgebraic, then there are $d \in \mathbb{N}$ and $M > 0$ such that $|g(x)| \leq x^d$ for all $x > M$.

3. Suppose a continuous function $g : \mathbb{R}^m \to \mathbb{R}$ satisfies $Q(x, g(x)) = 0$ for all $x \in \mathbb{R}^m$ and some nonzero polynomial $Q(X, T) \in \mathbb{R}[X, T]$, $X = (X_1, \ldots, X_m)$. Show that then g is semialgebraic.

Notes and comments

Classical sources for much of the material in this chapter are Koopman and Brown [36], where analytic varieties are locally decomposed into analytic cells, Whitney [63], which stratifies real algebraic sets into finitely many connected semialgebraic manifolds, and Łojasiewicz [40], where the Tarski-Seidenberg theorem is obtained by cell decomposition. See also the books by Benedetti and Risler [2] and by Bochnak, Coste and Roy [4].

The proof of (3.6) given here follows the proof of a corresponding local result in Łojasiewicz [40]. Other proofs are by Bochnak and Efroymson [5], Delzell [14], Coste and Roy, see [4]; for a short model-theoretic proof, see [17].

The result of exercise 2 in (3.7) means that the ordered field of real numbers is **polynomially bounded**. This property is shared by other o-minimal expansions of this structure, but of course not by $(\mathbb{R}, <, 0, 1, +, -, \cdot, \exp)$. Miller [43] discovered the surprising fact that every o-minimal expansion of the ordered field of reals is either polynomially bounded, or defines the exponential function. Polynomial boundedness implies the various Łojasiewicz inequalities, see [23].

CHAPTER 3

CELL DECOMPOSITION

Introduction

In this chapter we establish two important results in the subject of o-minimality: the monotonicity theorem (Section 1) and the cell decomposition theorem (Section 2). They are essential for everything that follows.

We work here with a fixed but arbitrary o-minimal structure $(R, <, \mathcal{S})$.

Instead of saying that a set $A \subseteq R^m$ belongs to $\mathcal{S}$ we will say that A is **definable**, as is usual in the literature on o-minimal structures. Similarly with maps.

For further conventions and notations, see Chapter 1, Sections 1, 2, 3.

§1. The monotonicity theorem and the finiteness lemma

(1.1) The monotonicity theorem describes definable one-variable functions.

(1.2) MONOTONICITY THEOREM. *Let $f:(a,b) \to R$ be a definable function on the interval (a,b). Then there are points $a_1 < \cdots < a_k$ in (a,b) such that on each subinterval (a_j, a_{j+1}), with $a_0 = a$, $a_{k+1} = b$, the function is either constant, or strictly monotone and continuous.*

(1.3) We derive this from the three lemmas below. In these lemmas we consider a definable function $f:I \to R$ on an interval I.

LEMMA 1. *There is a subinterval of I on which f is constant or injective.*

LEMMA 2. *If f is injective, then f is strictly monotone on a subinterval of I.*

LEMMA 3. *If f is strictly monotone, then f is continuous on a subinterval of I.*

(1.4) These lemmas imply the monotonicity theorem as follows:

Let

$$X := \{x \in (a,b) : \text{on some subinterval of } (a,b) \text{ containing } x \text{ the function } f \text{ is either constant, or strictly monotone and continuous}\}.$$

Now $(a,b) - X$ must be finite, since otherwise it would contain an interval I; applying successively lemmas 1, 2, and 3 we can make I so small that f is either constant, or strictly monotone and continuous, on I. But then $I \subseteq X$, a contradiction.

Since $(a,b) - X$ is finite, we can reduce the proof of the theorem to the case that $(a,b) = X$, by replacing (a,b) by each of the finitely many intervals of which the open set X consists. In particular, we may assume that f is continuous. By splitting up (a,b) further we can reduce to one of the following three cases.

CASE 1. *For all $x \in (a,b)$, f is constant on some neighborhood of x.*

CASE 2. *For all $x \in (a,b)$, f is strictly increasing on some neighborhood of x.*

CASE 3. *For all $x \in (a,b)$, f is strictly decreasing on some neighborhood of x.*

Case 1. Take $x_0 \in (a,b)$ and put

$$s := \sup\{x : x_0 < x < b, \ f \text{ is constant on } [x_0, x)\}.$$

Then $s = b$, since $s < b$ implies that f is constant on some neighborhood of s, contradiction. From $s = b$ it follows that f is constant on $[x_0, b)$. Similarly, we prove that f is constant on $(a, x_0]$. Therefore f is constant on (a,b).

Case 2. Take $x_0 \in (a,b)$ and put

$$s := \sup\{x : x_0 < x < b, \ f \text{ is strictly increasing on } [x_0, x)\}.$$

Then $s = b$, since $s < b$ leads to a contradiction as in case 1. Therefore f is strictly increasing on $[x_0, b)$. Similarly, f is strictly increasing on $(a, x_0]$. Hence f is strictly increasing on (a,b).

Case 3. This is handled in the same way as case 2.

(1.5) We now prove the lemmas.

PROOF OF LEMMA 1. If some $y \in R$ had infinite preimage $f^{-1}(y)$, then this preimage would contain a subinterval of I and f would take the constant value y on that subinterval. So we may assume that each $y \in R$ has finite preimage. Then $f(I)$ is infinite, and so contains an interval J. Define an "inverse" $g : J \to I$ by

$$g(y) := \min\{x \in I : f(x) = y\}.$$

Since g is injective by definition, $g(J)$ is infinite, and hence $g(J)$ contains a subinterval of I, and f is necessarily injective on this subinterval. $\square$

PROOF OF LEMMA 2. Let us write $I = (a,b)$. We assume here that f is injective and have to show that f is strictly monotone on some subinterval of I. For each $x \in I$ the interval (a,x) is a disjoint union of two subsets,

$$(a,x) = \{y \in (a,x) : f(y) < f(x)\} \cup \{y \in (a,x) : f(y) > f(x)\},$$

so one of the parts contains an interval (c,x), $a < c < x$. The interval (x,b) breaks up similarly. This shows that each $x \in I$ satisfies exactly one of the following four formulas:

$$\Phi_{++}(x) := \exists c_1, c_2 \in I \; [c_1 < x < c_2 \;\&\; \forall y \in (c_1,x) : f(y) > f(x) \\ \&\; \forall y \in (x,c_2) : f(y) > f(x)],$$

$$\Phi_{+-}(x) := \exists c_1, c_2 \in I \; [c_1 < x < c_2 \;\&\; \forall y \in (c_1,x) : f(y) > f(x) \\ \&\; \forall y \in (x,c_2) : f(y) < f(x)],$$

and $\Phi_{-+}(x)$ and $\Phi_{--}(x)$, which are defined similarly.

So I contains a subinterval all of whose points satisfy the same formula. Replacing I by this subinterval, and f by its restriction to that subinterval, we may assume that all points satisfy the same formula. This leads to four cases.

EASY CASE. *$\Phi_{-+}(x)$ for all x in I.*

For each x in I define $s(x) := \sup\{s \in (x,b) : \; f > f(x) \text{ on } (x,s]\}$. Then clearly $s(x) = b$, since $s(x) < b$ contradicts $\Phi_{-+}(s(x))$. Therefore f is strictly increasing on I.

The case that $\Phi_{+-}(x)$ for all x in I leads similarly to the conclusion that f is strictly decreasing on I.

DIFFICULT CASE. *$\Phi_{++}(x)$ for all x in I.*

Let $B := \{x \in I : \; \forall y \in I \; (y > x \text{ implies } f(y) > f(x))\}$. If B is infinite then B contains an interval, and on this interval f is strictly increasing, and we are done. So let us assume that B is finite. Passing to a subinterval to the right of all points of B we may assume

$$(*) \qquad \forall x \in I \; \exists y \in I \; (y > x \;\&\; f(y) < f(x)).$$

Let $c \in I$. We claim that for all large enough y in I we have $f(y) < f(c)$. Otherwise, we would have $f(y) > f(c)$ for all large enough $y \in (c,b)$. Take then $d \in [c,b)$ minimal such that $\forall y \; (d < y < b \text{ implies } f(y) > f(c))$.

If $f(d) > f(c)$ then d would not be minimal since $\Phi_{++}(d)$. So $f(d) < f(c)$. But by (*), there is e with $d < e < b$ and $f(e) < f(d)$, so $f(e) < f(c)$, contradiction. This proves the claim that $f(y) < f(c)$ for all large enough $y \in I$.

Define $y(c)$ as the least element of $[c, b)$ for which $f(y) < f(c)$ if $y(c) < y < b$. Note that $\Phi_{++}(c)$ gives $c < y(c)$ and $f(y(c)) < f(c)$. Minimality of $y(c)$ clearly implies that $y(c)$ satisfies the following formula $\Psi_{+-}(v)$:

$$\Psi_{+-}(v) := \exists v_1, v_2 \in I \left[v_1 < v < v_2 \ \&\ \forall z_1, z_2 \left(v_1 < z_1 < v < z_2 < v_2 \to f(z_1) > f(z_2)\right)\right].$$

Since c was arbitrary we have shown $\forall c \in I\ \exists v \in I\ \left(v > c \ \&\ \Psi_{+-}(v)\right)$.

Therefore $\Psi_{+-}(v)$ holds for all v in an interval of the form (d, b), $d \in I$. Replacing I by this subinterval we may as well assume that $\Psi_{+-}(v)$ holds on all of I.

A completely similar argument shows that we can pass to a still smaller subinterval on which Ψ_{-+} (defined in the obvious way) holds. But this is a contradiction since we cannot simultaneously have $\Psi_{+-}(v)$ and $\Psi_{-+}(v)$. The case that $\Phi_{--}(x)$ holds for all x in I is completely similar to the case we just handled. This finishes the proof of lemma 2. □

PROOF OF LEMMA 3. Let us assume that the strictly monotone function f is strictly increasing. (The case that f is strictly decreasing goes the same way.) Since $f(I)$ is infinite there is an interval $J \subseteq f(I)$. Take two points $r, s \in J$, $r < s$, and let c, d be their preimages: $f(c) = r$, $f(d) = s$, $c < d$. Clearly f defines an order preserving bijection of (c, d) onto (r, s). But the topology is defined in terms of the order, hence f is continuous on (c, d).

This finishes the proof of the lemmas and hence the proof of the monotonicity theorem is complete. □

Let us mention two easy but important consequences.

(1.6) COROLLARY 1. *Let $f : (a, b) \to R$ be definable. Then for each $c \in (a, b)$ the limits $\lim_{x \uparrow c} f(x)$ and $\lim_{x \downarrow c} f(x)$ exist in R_∞. Also the limits $\lim_{x \uparrow b} f(x)$ and $\lim_{x \downarrow a} f(x)$ exist in R_∞.*

COROLLARY 2. *Let $f : [a, b] \to R$ be continuous and definable. Then f takes a maximum and a minimum value on $[a, b]$.*

Here is the other basic result of this section.

(1.7) FINITENESS LEMMA. *Let $A \subseteq R^2$ be definable and suppose that for each $x \in R$ the fiber $A_x := \{y \in R : (x, y) \in A\}$ is finite. Then there is $N \in \mathbf{N}$ such that $|A_x| \leq N$ for all $x \in R$.*

PROOF. A point $(a,b)\in R^2$ will be called **normal** if there is a box $I\times J$ around (a,b) such that

either $(I\times J)\cap A=\emptyset$ (hence $(a,b)\notin A$),
or $(a,b)\in A$ and $(I\times J)\cap A=\Gamma(f)$ for some continuous function $f:I\to R$.

(Note that in the latter case f is necessarily unique and definable.) Also, a point $(a,-\infty)\in R\times R_\infty$ is called **normal** if there is a box $I\times J$ disjoint from A such that $a\in I$ and $J=(-\infty,b)$ for some b. Finally, $(a,+\infty)\in R\times R_\infty$ is called **normal** if there is a box $I\times J$ disjoint from A with $a\in I$ and $J=(b,+\infty)$ for some b. Note

(1) The sets

$$\{(a,b)\in R^2:\ (a,b)\text{ is normal}\},$$
$$\{a\in R:\ (a,-\infty)\text{ is normal}\},$$
$$\{a\in R:\ (a,+\infty)\text{ is normal}\},$$

are definable.

Next we define functions $f_1, f_2, \dots, f_n, \dots$ by

$$\operatorname{dom}(f_n) := \{x\in R:\ |A_x|\geq n\},$$
$$\text{and } f_n(x) := n^{\text{th}}\text{ element of } A_x.$$

Note that f_n is definable. (Its domain may of course be empty.)

Let $a\in R$ and take $n\geq 0$ maximal such that $f_1,f_2,\dots,f_n$ are defined and continuous on an interval containing a. We call the point a **good** or **bad**, according to whether

$$a\notin \operatorname{cl}(\operatorname{dom}(f_{n+1})) \quad \text{- “good”},$$
$$a\in \operatorname{cl}(\operatorname{dom}(f_{n+1})) \quad \text{- “bad”}.$$

Let $\mathcal{G}$ be the set of good points and $\mathcal{B}$ the set of bad points. Note that if $a\in\mathcal{G}$ then (with n as above) the domain of f_{n+1} is disjoint from an entire interval around a on which $f_1, f_2, \dots, f_n$ are defined and continuous. This shows that for $a\in\mathcal{G}$ we have

(2) *$|A_x|$ is constant on an interval around a.*

(3) *(a,b) is normal for all $b\in R_\infty$.*

The key point of the proof is to show that $\mathcal{B}$ and $\mathcal{G}$ are *definable* sets. (This is not clear from their definitions which involve a parameter $n\in\mathbf{N}$ depending on a.) To this end we shall show

(4) *If $a\in\mathcal{B}$ then there is a least $b\in R_\infty$ such that (a,b) is not normal.*

To see this, let us introduce for $a \in \mathcal{B}$ the elements $\lambda(a,-)$, $\lambda(a,0)$, $\lambda(a,+)$ of R_∞, where n has the same meaning as before:

$$\begin{aligned}
\lambda(a,-) &:= \lim_{x \uparrow a} f_{n+1}(x) \text{ if } f_{n+1} \text{ is defined on some interval } (t,a),\\
&:= +\infty \text{ otherwise,}\\
\lambda(a,0) &:= f_{n+1}(a) \text{ if } a \in \operatorname{dom}(f_{n+1}),\\
&:= +\infty \text{ otherwise,}\\
\lambda(a,+) &:= \lim_{x \downarrow a} f_{n+1}(x) \text{ if } f_{n+1} \text{ is defined on some interval } (a,t),\\
&:= +\infty \text{ otherwise.}
\end{aligned}$$

Now let $\beta(a) := \min\{\lambda(a,-),\ \lambda(a,0),\ \lambda(a,+)\}$. Then by checking the various possibilities it is easy to see that $\beta(a)$ is the least $b \in R_\infty$ such that (a,b) is not normal. This proves claim (4), and together with (1) and (3) it implies that $\mathcal{B}$ and $\mathcal{G}$ are definable sets.

The rest of the proof is straightforward. Suppose first that $\mathcal{B}$ is finite, say

$$\mathcal{B} = \{a_1,\ldots,a_k\}, \text{ with } -\infty = a_0 < a_1 < \cdots < a_k < a_{k+1} = +\infty.$$

We claim that $|A_x|$ is constant on each interval (a_i,a_{i+1}); just take any point a in this interval, and let $n = |A_a|$. Then by (2) the set $\{x \in (a_i,a_{i+1})\colon\ |A_x| = n\}$ is open, and for the same reason the set $\{x \in (a_i,a_{i+1}) : |A_x| \neq n\}$ is open. Since both sets are definable the latter set must be empty.

Suppose now that $\mathcal{B}$ is not finite. We shall derive a contradiction from this assumption, and that will finish the proof. Recall that $\beta(a)$ is, for $a \in \mathcal{B}$, the least $b \in R_\infty$ such that (a,b) is not normal. Define the sets

$$\begin{aligned}
\mathcal{B}_- &:= \{a \in \mathcal{B}\colon\ \exists y\ (y < \beta(a)\ \&\ (a,y) \in A)\},\\
\mathcal{B}_+ &:= \{a \in \mathcal{B}\colon\ \exists y\ (y > \beta(a)\ \&\ (a,y) \in A)\},
\end{aligned}$$

and the functions $\beta_-\colon \mathcal{B}_- \to R$ and $\beta_+\colon \mathcal{B}_+ \to R$ by

$$\begin{aligned}
\beta_-(a) &:= \max\{y\colon\ y < \beta(a)\ \&\ (a,y) \in A\},\\
\beta_+(a) &:= \min\{y\colon\ y > \beta(a)\ \&\ (a,y) \in A\}.
\end{aligned}$$

Since $\mathcal{B}$ is infinite by assumption, one of the sets $\mathcal{B}_- \cap \mathcal{B}_+$, $\mathcal{B}_- - \mathcal{B}_+$, $\mathcal{B}_+ - \mathcal{B}_-$, $\mathcal{B} - (\mathcal{B}_- \cup \mathcal{B}_+)$ is infinite, and each of these four cases leads to a contradiction. We only show this in the case that $\mathcal{B}_- \cap \mathcal{B}_+$ is infinite. (The other cases are similar.) Since β_-, β, and β_+ are definable functions, there is by the monotonicity theorem an interval $I \subseteq \mathcal{B}_- \cap \mathcal{B}_+$ on which each of the functions β_-, β, β_+ is continuous. Note that $\beta_- < \beta < \beta_+$ on I. Now I splits into two subsets:

$$\begin{aligned}
&\{x \in I\colon\ (x,\beta(x)) \in A\},\\
\text{and } &\{x \in I\colon\ (x,\beta(x)) \notin A\},
\end{aligned}$$

and one of these subsets contains an interval. Replacing I by this subinterval we may assume that either $\Gamma(\beta|I) \subseteq A$, or $\Gamma(\beta|I) \cap A = \emptyset$. In either case it is clear that $\Gamma(\beta|I)$ consists of normal points, since β_-, β, β_+ are continuous on I. Now we have a contradiction, since $(a, \beta(a))$ is never normal. □

(1.8) We now combine the finiteness lemma with the monotonicity theorem in the following result to be used in the next section.

Let $A \subseteq R^2$ be definable such that A_x is finite for each $x \in R$. Then there are points $a_1 < \cdots < a_k$ in R such that the intersection of A with each vertical strip $(a_i, a_{i+1}) \times R$ has the form $\Gamma(f_{i1}) \cup \cdots \cup \Gamma(f_{in(i)})$ for certain definable continuous functions $f_{ij} : (a_i, a_{i+1}) \to R$ with $f_{i1}(x) < \cdots < f_{in(i)}(x)$ for $x \in (a_i, a_{i+1})$. (Here we have set $a_0 := -\infty$, $a_{k+1} := +\infty$.)

(1.9) EXERCISES.

1. Suppose the function $f:(a,b) \to R$ on the interval (a,b) is definable. Show there exist elements $a_1, \ldots, a_k$ with the property of the monotonicity theorem such that $a_1, \ldots, a_k$ are definable in the model-theoretic structure $(R, <, \Gamma(f))$.

2. Let I and J be intervals and $f:I \to R$ and $g:J \to R$ strictly monotone definable functions such that $f(I) \subseteq g(J)$ and $\lim_{s \to r(I)} f(s) = \lim_{t \to r(J)} g(t)$ in R_∞, where $r(I)$ and $r(J)$ are the right endpoints of the intervals I and J in R_∞. Show that near these right endpoints f and g are reparametrizations of each other, that is, there are subintervals I' of I and J' of J, with $r(I) = r(I')$, $r(J) = r(J')$ and a strictly increasing definable bijection $h : I' \to J'$ such that $f(s) = g(h(s))$ for all $s \in I'$. (This will be used in Chapter 6, (4.3).)

§2. The cell decomposition theorem

(2.1) In this section we show that a definable subset of R^m splits into finitely many cells—definable sets of an especially simple form—and that each definable function on a subset of R^m is "cellwise" continuous. For $m = 1$ this reduces for *sets* to the definition of "o-minimal" and for *functions* to the monotonicity theorem of the last section.

(2.2) For each definable set X in R^m we put

$$C(X) := \{f : X \to R : f \text{ is definable and continuous}\},$$
$$C_\infty(X) := C(X) \cup \{-\infty, +\infty\},$$

where we regard $-\infty$ and $+\infty$ as constant functions on X.

For f, g in $C_\infty(X)$ we write $f < g$ if $f(x) < g(x)$ for all $x \in X$, and in this case we put

$$(f, g)_X := \{(x, r) \in X \times R : f(x) < r < g(x)\}.$$

So $(f,g)_X$ is a definable subset of R^{m+1}.

Usually we write just (f,g) instead of $(f,g)_X$ when X is clear from the context.

(2.3) DEFINITION. Let $(i_1,\ldots,i_m)$ be a sequence of zeros and ones of length m. An $(i_1,\ldots,i_m)$-**cell** is a definable subset of R^m obtained by induction on m as follows:

(i) a (0)-cell is a one-element set $\{r\} \subseteq R$ (a "point"), a (1)-cell is an interval $(a,b) \subseteq R$;
(ii) suppose $(i_1,\ldots,i_m)$-cells are already defined; then an $(i_1,\ldots,i_m,0)$-cell is the graph $\Gamma(f)$ of a function $f \in C(X)$, where X is an $(i_1,\ldots,i_m)$-cell; further, an $(i_1,\ldots,i_m,1)$-cell is a set $(f,g)_X$ where X is an $(i_1,\ldots,i_m)$-cell and $f,g \in C_\infty(X)$, $f < g$. (See figure.)

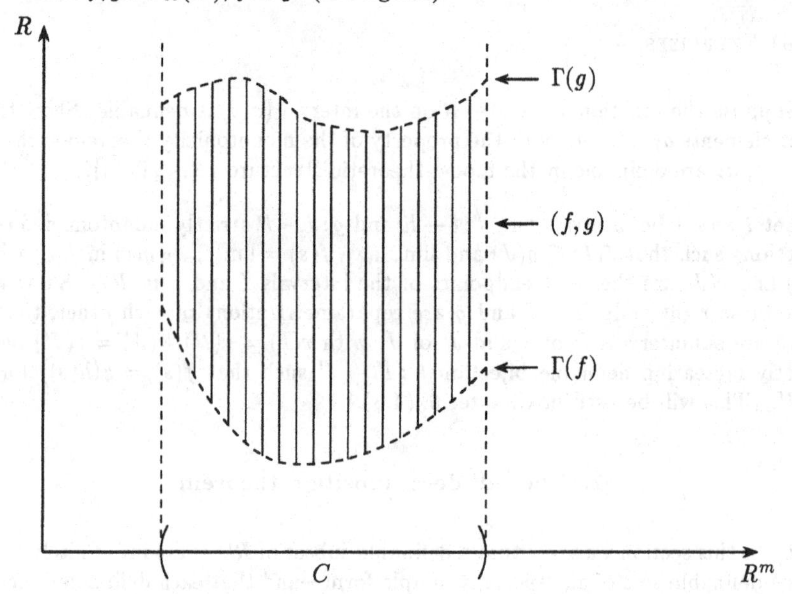

So a $(0,0)$-cell is a "point" $\{(r,s)\} \subseteq R^2$, a $(0,1)$-cell is an "interval" on a vertical line $\{a\} \times R$, and a $(1,0)$-cell is the graph of a continuous definable function defined on an interval. Note that a box in R^m is a $(1,\ldots,1)$-cell.

(2.4) TERMINOLOGY. *A* **cell in** R^m *is an* $(i_1,\ldots,i_m)$*-cell, for some (necessarily unique) sequence* $(i_1,\ldots,i_m)$. *Since the* $(1,\ldots,1)$*-cells are exactly the cells which are open in their ambient space* R^m *we call these* **open cells**.

(2.5) The non-open cells are "thin":

The union of finitely many non-open cells in R^m *has empty interior.* This is easily checked and we shall use this fact later on without further mention. Here is another topological fact:

Each cell is locally closed, i.e. open in its closure.

To see this, let $C \subseteq R^{m+1}$ be a cell. Put $B := \pi(C) \subseteq R^m$ and assume inductively that the cell B is open in its closure $\mathrm{cl}(B)$, so that $\mathrm{cl}(B) - B$ is a closed set. If $C = \Gamma(f)$ with $f: B \to R$ a definable continuous function, then $\mathrm{cl}(C) - C$ is contained in $(\mathrm{cl}(B) - B) \times R$, hence C is open in the closed set $C \cup \big((\mathrm{cl}(B) - B) \times R\big)$. If $C = (f, g)$ with $f, g: B \to R$ definable continuous functions on B, $f < g$, then one easily verifies that $\mathrm{cl}(C) - C$ is contained in $\Gamma(f) \cup \Gamma(g) \cup \big((\mathrm{cl}(B) - B) \times R\big)$ and that C is open in the closed set $C \cup \Gamma(f) \cup \Gamma(g) \cup \big((\mathrm{cl}(B) - B) \times R\big)$. (Draw a picture.) The other cases are done similarly.

(2.6) CONVENTION. *We also consider the one-point space R^0 as a cell, more precisely, as a ( )-cell, where ( ) is the sequence of length* 0. *(It is an open cell; there are no other ( )-cells.)*

(In this way clause (i) in (2.3) appears as the case $m = 0$ of clause (ii). It allows us to start inductions with $m = 0$.)

(2.7) Each cell is homeomorphic under a coordinate projection to an open cell. We now make this explicit. Let $i = (i_1, \dots, i_m)$ be a sequence of zeros and ones.

Define $p_i : R^m \to R^k$ as follows: let $\lambda(1) < \dots < \lambda(k)$ be the indices $\lambda \in \{1, \dots, m\}$ for which $i_\lambda = 1$, so that $k = i_1 + \dots + i_m$; then

$$p_i(x_1, \dots, x_m) := \big(x_{\lambda(1)}, \dots, x_{\lambda(k)}\big).$$

It is easy to show by induction on m that p_i maps each i-cell A homeomorphically onto an open cell $p_i(A)$ in R^k. We denote $p_i(A)$ also by $p(A)$ and the homeomorphism $p_i|A : A \to p(A)$ by p_A. Clearly $p_A = id_A$ if A is an open cell.

(2.8) If A is a cell in R^{m+1} then $\pi(A)$ is a cell in R^m, where $\pi : R^{m+1} \to R^m$ is the projection on the first m coordinates. Here is a simple application of this fact.

(2.9) PROPOSITION. *Each cell is definably connected.*

PROOF. For intervals and points this is stated in Chapter 1, (3.6). If A is a cell in R^{m+1}, then we assume inductively that the cell $\pi(A)$ in R^m is definably connected and use the fact that each fiber $\pi^{-1}(x) \cap A$ is definably connected. □

(2.10) DEFINITION. A **decomposition** of R^m is a special kind of partition of R^m into finitely many cells. The definition is by induction on m:

(i) a decomposition of $R^1 = R$ is a collection

$$\{(-\infty, a_1), (a_1, a_2), \dots, (a_k, +\infty), \{a_1\}, \dots, \{a_k\}\}$$

where $a_1 < \cdots < a_k$ are points in R;

(ii) a decomposition of R^{m+1} is a finite partition of R^{m+1} into cells A such that the set of projections $\pi(A)$ is a decomposition of R^m. (Here $\pi : R^{m+1} \to R^m$ is the usual projection map.)

Let $\mathcal{D} = \{A(1), \ldots, A(k)\}$ be a decomposition of R^m, $A(i) \neq A(j)$ if $i \neq j$, and let for each $i \in \{1, \ldots, k\}$ functions $f_{i1} < \cdots < f_{in(i)}$ in $C(A_i)$ be given.

Then

$$\mathcal{D}_i := \{(-\infty, f_{i1}), (f_{i1}, f_{i2}), \ldots, (f_{in(i)}, +\infty), \Gamma(f_{i1}), \ldots, \Gamma(f_{in(i)})\}$$

is a partition of $A(i) \times R$ and one easily checks that $\mathcal{D}^* := \mathcal{D}_1 \cup \cdots \cup \mathcal{D}_k$ is a decomposition of R^{m+1}, and that every decomposition of R^{m+1} arises in this way from a decomposition $\mathcal{D}$ of R^m. We write $\mathcal{D} = \pi(\mathcal{D}^*)$. (See figure.)

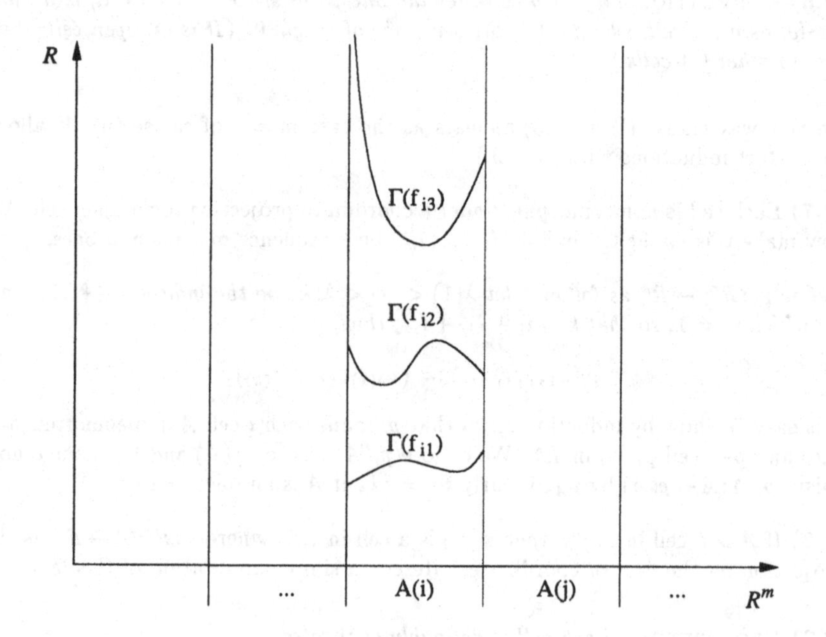

A decomposition $\mathcal{D}$ of R^m is said to **partition** a set $S \subseteq R^m$ if each cell in $\mathcal{D}$ is either part of S or disjoint from S, in other words, if S is a union of cells in $\mathcal{D}$. We are now ready to state the main result of this chapter.

(2.11) CELL DECOMPOSITION THEOREM.

(I_m) *Given any definable sets $A_1, \ldots, A_k \subseteq R^m$ there is a decomposition of R^m partitioning each of $A_1, \ldots, A_k$.*

(II_m) *For each definable function $f : A \to R$, $A \subseteq R^m$, there is a decomposition $\mathcal{D}$ of R^m partitioning A such that the restriction $f|B : B \to R$ to each cell $B \in \mathcal{D}$ with $B \subseteq A$ is continuous.*

(2.12) The proof is by induction on m. Note that (I_1) holds by o-minimality, and that (II_1) follows easily from the monotonicity theorem.

We now assume that (I_1), ..., (I_m) and (II_1), ..., (II_m) hold, and shall derive first (I_{m+1}) and then (II_{m+1}).

The proof is lengthy. The first step is to generalize the finiteness lemma of the previous section. Call a set $Y \subseteq R^{m+1}$ **finite over** R^m if for each $x \in R^m$ the fiber $Y_x := \{r \in R : (x,r) \in Y\}$ is finite; call Y **uniformly finite over** R^m if there is $N \in \mathbf{N}$ such that $|Y_x| \le N$ for all $x \in R^m$. We shall use the finiteness lemma and the inductive assumption to prove first

(2.13) Lemma (uniform finiteness property). *Suppose the definable subset Y of R^{m+1} is finite over R^m. Then Y is uniformly finite over R^m.*

Proof. A box $B \subseteq R^m$ will be called Y**-good** if for each point $(x,r) \in Y$ with $x \in B$ there is an interval I around r such that $Y \cap (B \times I) = \Gamma(f)$ for some continuous function $f: B \to R$. (Note: this f is then uniquely determined by Y, B, and I, and is definable.)

Claim 1. Suppose the box $B \subseteq R^m$ is Y-good; then there are continuous definable functions $f_1 < \cdots < f_k$ in $C(B)$ such that $Y \cap (B \times R) = \Gamma(f_1) \cup \cdots \cup \Gamma(f_k)$.

To see this, let us fix $x \in B$ and write $Y_x = \{r_1, \ldots, r_k\}$ with $r_1 < \cdots < r_k$. Take intervals $I_1, \ldots, I_k$ around $r_1, \ldots, r_k$ respectively, and continuous functions $f_1, \ldots, f_k: B \to R$ such that $Y \cap (B \times I_j) = \Gamma(f_j)$, $j = 1, \ldots, k$.

Subclaim a. $f_1 < \cdots < f_k$.

Let us prove only $f_1 < f_2$, the other inequalities following in the same way. Suppose there is a point $p \in B$ with $f_1(p) = f_2(p)$. So $f_2(p) \in I_1$, and by continuity of f_2 there is a neighborhood $U \subseteq B$ of p such that $f_2(U) \subseteq I_1$. Since $Y \cap (U \times I_1) = \Gamma(f_1|U)$ and $\Gamma(f_2|U) \subseteq Y \cap (U \times I_1)$ it follows that $f_1|U = f_2|U$. This argument shows that the set $\{p \in B : f_1(p) = f_2(p)\}$ is open. Since $\{p \in B : f_1(p) < f_2(p)\}$ and $\{p \in B : f_1(p) > f_2(p)\}$ are also open and B is definably connected, by (2.9), while $f_1(x) = r_1 < r_2 = f_2(x)$, it follows that $f_1 < f_2$.

Subclaim b. $Y \cap (B \times R) = \Gamma(f_1) \cup \cdots \cup \Gamma(f_k)$.

Take any point $(a,s) \in Y \cap (B \times R)$ and let $f: B \to R$ be a continuous definable function such that $f(a) = s$ and $\Gamma(f) \subseteq Y$. Since $(x, f(x)) \in Y$ it follows that $f(x) = r_i = f_i(x)$ for some $i \in \{1, \ldots, k\}$. As in the proof of subclaim a we obtain from this $f = f_i$. This finishes the proof of claim 1.

A point $x \in R^m$ will be called Y**-good** if x belongs to a Y-good box. Note that the set of Y-good points is definable.

Claim 2. If $A \subseteq R^m$ is a definably connected set and all points of A are Y-good, then there are continuous functions $f_1 < \cdots < f_k$ in $C(A)$ such that

$$Y \cap (A \times R) = \Gamma(f_1) \cup \cdots \cup \Gamma(f_k).$$

To prove this we choose a point $x \in A$, if A is nonempty, and let $k = |Y_x|$. By claim 1 the set $\{a \in A : |Y_a| = k\}$ is open and closed in A, hence $|Y_a| = k$ for all $a \in A$. Again, by claim 1, it is clear that $f_1, \ldots, f_k$ are continuous.

Claim 3. Each open cell in R^m contains a Y-good point.

It is of course enough to show that each box B in R^m contains a Y-good point. Write

$$B = B' \times (a,b), \ \ B' \text{ a box in } R^{m-1}.$$

For each point $p \in B'$ consider the set

$$Y(p) := \{(r,s) \in R^2 : a < r < b \ \& \ (p,r,s) \in Y\},$$

which is finite over R. Now we apply (1.8) to $A = Y(p)$ and conclude that the set $\{r \in R : r \text{ is not } Y(p)\text{-good}\}$ is finite. Therefore, the definable set

$$\text{Bad}(Y) := \{(p,r) \in B : r \text{ is not } Y(p)\text{-good}\}$$

has no interior point. By the inductive assumption (I_m) there is a decomposition of R^m which partitions B and $\text{Bad}(Y)$. Take an open cell C of this partition such that $C \subseteq B$. Then $C \cap \text{Bad}(Y) = \emptyset$, so if we replace B by a box contained in C we have reduced to the case that $\text{Bad}(Y) = \emptyset$, i.e., for each $p \in B'$ we can apply claim 2 (with $Y(p) \subseteq R^2$ instead of Y) to find a number $k(p) \in \mathbf{N}$ such that $|Y_x| = k(p)$ for each point $x = (p,r) \in B$. Next we have to show that there is a finite bound on the numbers $k(p)$, $p \in B'$.

To this end, we choose an $r \in (a,b)$ and consider the set

$$Y^r := \{(p,s) : (p,r,s) \in Y\} \subseteq R^m.$$

Since Y is finite over R^m, the set Y^r is finite over R^{m-1}, so by the inductive assumption Y^r is uniformly finite over R^{m-1}, i.e. there is $N \in \mathbf{N}$ such that for each $p \in B'$: $|\{s \in R : (p,s) \in Y^r\}| \leq N$, that is, $|Y_{(p,r)}| \leq N$ for all $p \in B'$. Hence $k(p) \leq N$ for all $p \in B'$. Thus $|Y_x| \leq N$ for all $x \in B$.

For each $i \in \{0,\ldots,N\}$ let $B_i := \{x \in B : |Y_x| = i\}$, and define the functions $f_{i1}, \ldots, f_{ii}$ on B_i by $f_{i1}(x) < \cdots < f_{ii}(x)$, and $Y_x = \{f_{i1}(x), \ldots, f_{ii}(x)\}$. Applying the inductive assumption (II_m) to each f_{ij} separately, and then using (I_m) to find a common refinement of the decompositions obtained via (II_m), we get a decomposition $\mathcal{D}$ of R^m partitioning each of the sets B_i, such that for each $A \in \mathcal{D}$, if $A \subseteq B_i$, then $f_{ij}|A$ is continuous, $j = 0, \ldots, i$. Since B is open and is partitioned by $\mathcal{D}$, there is an open cell $A \in \mathcal{D}$ with $A \subseteq B$. Now $B = \bigcup_i B_i$, so $A \subseteq B_i$ for some i, therefore the functions $f_{i1}, \ldots, f_{ij}$ are continuous on A. Hence each point of A is Y-good. Since $A \subseteq B$ this establishes claim 3.

The proof of the lemma now proceeds as follows. Take a decomposition $\mathcal{D}$ of R^m partitioning the set of Y-good points. Let $A \in \mathcal{D}$. If A is open, then by claim 3 the cell A contains a Y-good point, so all points of A are Y-good. By claim 2 there is then a number $N_A \in \mathbf{N}$ such that $|Y_x| \leq N_A$ for all $x \in A$. By an easy exercise using the definable homeomorphism p_A (see (2.7)) such a number N_A also exists for the non-open cells $A \in \mathcal{D}$. Now take $N := \max\{N_A : A \in \mathcal{D}\}$. Then $|Y_x| \leq N$ for all x in R^m. □

(2.14) Before we proceed with the proof of (I_{m+1}) we recall from Chapter 1, (3.3) that a definable set $S \subseteq R$ has finite boundary $\mathrm{bd}(S)$, and that the interval between two successive boundary points is either part of S or disjoint from S.

For a definable set $A \subseteq R^{m+1}$ we put

$$\mathrm{bd}_m(A) := \{(x, r) \in R^{m+1} : r \in \mathrm{bd}(A_x)\},$$

and we note that $\mathrm{bd}_m(A)$ is a definable set which is finite over R^m, so that we can apply the uniform finiteness property. This is the idea of the proof below.

(2.15) Proof of (I_{m+1}). Let $A_1, \ldots, A_k$ be definable subsets of R^{m+1}. Put

$$Y := \mathrm{bd}_m(A_1) \cup \cdots \cup \mathrm{bd}_m(A_k).$$

Then $Y \subseteq R^{m+1}$ is definable and finite over R^m, so there is $M \in \mathbf{N}$ such that $|Y_x| \leq M$ for all x in R^m. For each $i \in \{0, \ldots, M\}$ let $B_i := \{x \in R^m : |Y_x| = i\}$, and define functions $f_{i1}, f_{i2}, \ldots, f_{ii}$ on B_i by

$$Y_x = \{f_{i1}(x), \ldots, f_{ii}(x)\}, \; f_{i1}(x) < \cdots < f_{ii}(x).$$

Further, put $f_{i0} := -\infty$, $f_{ii+1} := +\infty$ (functions on B_i). Finally we define for each $\lambda \in \{1, \ldots, k\}$, $i \in \{0, \ldots, M\}$ and $1 \leq j \leq i$

$$C_{\lambda ij} := \{x \in B_i : f_{ij}(x) \in (A_\lambda)_x\},$$

and for each $\lambda \in \{1, \ldots, k\}$, $i \in \{0, \ldots, M\}$ and $0 \leq j \leq i$

$$D_{\lambda ij} := \{x \in B_i : (f_{ij}(x), f_{ij+1}(x)) \subseteq (A_\lambda)_x\}.$$

We now take a decomposition $\mathcal{D}$ of R^m which partitions each set B_i, each set $C_{\lambda ij}$, each set $D_{\lambda ij}$, and which has moreover the following property:

if $E \in \mathcal{D}$ is contained in B_i, then $f_{i1}|E, \ldots, f_{ii}|E$ are continuous functions.

(Such a decomposition exists by the inductive assumptions (I_m) and (II_m).)

For each cell $E \in \mathcal{D}$ we let $\mathcal{D}_E$ be the following partition of $E \times R$:

$$\mathcal{D}_E := \{(f_{i0}|E, f_{i1}|E), \ldots, (f_{ii}|E, f_{ii+1}|E), \Gamma(f_{i1}|E), \ldots, \Gamma(f_{ii}|E)\},$$

where $i \in \{0, \ldots, M\}$ is such that $E \subseteq B_i$. Then $\mathcal{D}^* := \bigcup\{\mathcal{D}_E : E \in \mathcal{D}\}$ is a decomposition of R^{m+1} which partitions each set $A_1, \ldots, A_k$. □

The proof of (II_{m+1}) will be based on the following elementary lemma.

(2.16) LEMMA. *Let X be a topological space, $(R_1,<)$, $(R_2,<)$ dense linear orderings without endpoints and $f : X \times R_1 \to R_2$ a function such that for each $(x,r) \in X \times R_1$*

(i) *$f(x,\cdot): R_1 \to R_2$ is continuous and monotone on R_1,*
(ii) *$f(\cdot,r): X \to R_2$ is continuous at x.*

Then f is continuous.

PROOF. Let $(x,r) \in X \times R_1$ and $f(x,r) \in J$, where J is an interval in R_2. We shall find a neighborhood U of x and an interval I around r such that $f(U \times I) \subseteq J$. By (i) there are r_-, r_+ in R_1 such that $r_- < r < r_+$ and $f(x,r_-), f(x,r_+) \in J$. Now use (ii) to get a neighborhood U of x such that $f(U \times \{r_-\}) \subseteq J$ and $f(U \times \{r_+\}) \subseteq J$. We claim that then $f(U \times I) \subseteq J$ for $I = (r_-, r_+)$.

Let $x' \in U$ and $r_- < r' < r_+$. Assume $f(x',\cdot)$ is increasing. (The case that $f(x',\cdot)$ is decreasing goes the same way.) Then $f(x',r_-) \leq f(x',r') \leq f(x',r_+)$ and $f(x',r_-)$, $f(x',r_+)$ are both in J, hence $f(x',r')$ is in J. □

(2.17) PROOF OF (II_{m+1}). Let $f: A \to R$ be a definable function on a definable set $A \subseteq R^{m+1}$. We have to show that f is "cellwise" continuous. Because of (I_{m+1}) it suffices to show

$$(*) \qquad \begin{cases} A \text{ can be partitioned into finitely many definable sets } A_1, \ldots, A_k \\ \text{such that } f|A_i : A_i \to R \text{ is continuous for } i = 1, \ldots, k. \end{cases}$$

Again because of (I_{m+1}) the set A allows a partition into finitely many cells. So in order to prove $(*)$ we may assume that A is already a cell. If the cell A is not open in R^{m+1}, we use the definable homeomorphism $p_A: A \to p(A)$. Since $p(A) \subseteq R^n$ for some $n \leq m$, it follows from the inductive assumption (II_n) that the set $p(A)$ can be partitioned into definable sets $B_1, \ldots, B_k$ such that $(f \circ p_A^{-1})|B_j$ is continuous for each j. Hence A is partitioned into $p_A^{-1}(B_1), \ldots, p_A^{-1}(B_k)$, and the restriction of f to each of these sets is continuous.

This establishes $(*)$ for A non-open. Suppose now that A is an open cell. Call f **well-behaved** at a point $(p,r) \in A$ if $p \in C$ for some box $C \subseteq R^m$ and $a < r < b$ for some a, b in R such that

(i) $C \times (a,b)$ is contained in A,
(ii) for all $x \in C$ the function $f(x,\cdot)$ is continuous and monotone on (a,b),
(iii) the function $f(\cdot,r)$ is continuous at p.

Let A^* be the set of all points of A at which f is well-behaved. Note that A^* is definable.

Claim. A^* is dense in A.

To prove this it suffices to show that, given any box $B \subseteq R^m$ and $-\infty < a < c < +\infty$ such that $B \times (a,c)$ is contained in A, the box $B \times (a,c)$ intersects A^*. Now there is by the monotonicity theorem for each $x \in B$ a largest $\lambda(x) \in (a, c]$ such that the one-variable function $f(x, \cdot)$ is continuous and monotone on $(a, \lambda(x))$. Since $\lambda : B \to R$ is definable, there is by (II_m) a box $C \subseteq B$ on which λ is continuous. Taking C small enough we may assume that $b \leq \lambda(x)$ for all $x \in C$ and a fixed $b \in (a, c)$. Choose any element r in (a, b). The function $f(\cdot, r) : C \to R$ is continuous on some smaller box, by (II_m). Replacing C by this smaller box, we see that f is well-behaved at each point (p, r) with p in C. This establishes the claim.

Now we take a decomposition $\mathcal{D}$ of R^{m+1} that partitions both A and A^*. (Such a decomposition exists by (I_{m+1}).) Let $D \in \mathcal{D}$ be any open cell contained in A.

We need only show that f is continuous on D. From $D \subseteq A$ we obtain $D \subseteq A^*$, since D intersects A^* by the claim, and $\mathcal{D}$ partitions A^*. In particular for each point (p, r) in D the function $f(\cdot, r)$ is continuous at p. Therefore D is the union of boxes $C \times (a, b)$ satisfying the conditions (i), (ii), and (iii) above, for each point $p \in C$, $a < r < b$. By lemma (2.16) the function f is continuous on each such box, hence f is continuous on D. This concludes the proof of the cell decomposition theorem. □

We finish this section with an application of cell decomposition to definably connected components, and some exercises.

A **definably connected component** of a nonempty definable set $X \subseteq R^m$ is by definition a maximal definably connected subset of X.

(2.18) PROPOSITION. *Let $X \subseteq R^m$ be a nonempty definable set. Then X has only finitely many definably connected components. They are open and closed in X and form a finite partition of X.*

PROOF. Let $\{C_1, \ldots, C_k\}$ be a partition of X into k disjoint cells. For each nonempty set of indices $I \subseteq \{1, \ldots, k\}$, put $C_I := \bigcup_{i \in I} C_i$. Among the $2^k - 1$ sets C_I, let C' be maximal with respect to being definably connected. We claim

$$\text{If a set } Y \subseteq X \text{ is definably connected and } C' \cap Y \neq \emptyset, \text{ then } Y \subseteq C'.$$

To see why, put $C_Y := \bigcup\{C_i : C_i \cap Y \neq \emptyset\}$. Since the C_i's cover X we have $Y \subseteq C_Y$, so C_Y is the union of Y with certain cells that intersect Y. Hence C_Y is definably connected. But $C' \cap C_Y$ contains the nonempty set $C' \cap Y$, so $C' \cup C_Y$ is definably connected. By maximality of C' it follows that $C' \cup C_Y = C'$. Hence $Y \subseteq C_Y \subseteq C'$, which proves the claim. It follows in particular that C' is a definably connected component of X. It also follows that the sets of the form C' form a (finite) partition of X. Further the claim shows that the sets C' are the only definably connected components of X. Note that because the closure in X of a definably connected subset of X is also definably connected, the definably connected components of X are closed in X. Hence they are also open in X. □

(2.19) EXERCISES.

In the first four exercises we strengthen the cell decomposition theorem in several ways. This requires a somewhat technical notion of regularity:

(i) Call an open cell $C \subseteq R^m$ **regular** if for each $i \in \{1,\ldots,m\}$ and each two points $x, y \in C$ that differ only in the ith coordinate and each point $z \in R^m$ that differs from x and y only in the ith coordinate, we have $x_i < z_i < y_i \Rightarrow z \in C$. (Example: boxes in R^m are regular.)

(ii) Consider a definable function $f: C \to R$, where $C \subseteq R^m$ is a regular open cell. Call f **strictly increasing in the** i^{th} **coordinate** ($1 \le i \le m$) if for all points x, y in C that differ only in the ith coordinate, with $x_i < y_i$, we have $f(x) < f(y)$; similarly we define the notions of "f is **strictly decreasing in the** i^{th} **coordinate**" and "f **is independent of the** i^{th} **coordinate**", the latter meaning that $f(x) = f(y)$ whenever $x, y \in C$ differ only in the ith coordinate. Finally, let us say that f is **regular** if f is continuous, and for each $i \in \{1,\ldots,m\}$ the function f is either strictly increasing in the ith coordinate, or strictly decreasing in the ith coordinate, or independent of the ith coordinate. (Which of these three cases takes holds may depend on i.)

1. Let $C \subseteq R^m$ be a regular open cell and $f: C \to R$ a regular definable function. Show that the open cells $(-\infty, f)$, $(f, +\infty)$ and $C \times R = (-\infty, +\infty)$ in R^{m+1} are regular. Show that if $g: C \to R$ is a second regular definable function with $f < g$, then the open cell (f, g) in R^{m+1} is regular.

2. Prove by induction on m the **regular cell decomposition theorem:**

(I_m) For any definable sets $A_1, \ldots, A_k \subseteq R^m$ there is a decomposition of R^m partitioning each A_i, all of whose open cells are regular.

(II_m) For each definable function $f: A \to R$, $A \subseteq R^m$, there is a decomposition $\mathcal{D}$ of R^m partitioning A all of whose open cells are regular, and such that for each open cell $C \in \mathcal{D}$ with $C \subseteq A$ the restriction $f|C$ is regular.

3. Let C be a cell in R^m, $D = (\alpha, \beta)_C$ a cell in R^{m+1} and $f: D \to R$ a definable function such that for all $x \in C$ the function $f(x, \cdot): (\alpha(x), \beta(x)) \to R$ is continuous. Show that C can be partitioned into cells $C_1, \ldots, C_k$ such that, with $\alpha_i := \alpha|C_i$, $\beta_i := \beta|C_i$, each restriction $f|(\alpha_i, \beta_i): (\alpha_i, \beta_i) \to R$ is continuous.

4. Improve the cell decomposition theorem as follows:

(I_m) If the sets $A_1, \ldots, A_k \subseteq R^m$ are definable, then there is a decomposition of R^m partitioning each set A_i, all of whose cells are definable in the model-theoretic structure $(R, <, A_1, \ldots, A_k)$.

(II_m) Let the function $f: A \to R$, $A \subseteq R^m$, be definable. Then there is a decomposition $\mathcal{D}$ of R^m partitioning A, such that the restriction $f|B$ to each cell $B \in \mathcal{D}$ with $B \subseteq A$ is continuous, and each cell in $\mathcal{D}$ is definable in the model-theoretic structure $(R, <, \Gamma(f))$.

5. Let $X_1, \ldots, X_k \subseteq R^m$ be distinct nonempty definably connected sets and X their union. Define a graph with vertex set $\{X_1, \ldots, X_k\}$ by putting an edge between X_i and X_j ($i \neq j$) if $X_i \cap \mathrm{cl}(X_j) \neq \emptyset$ or $\mathrm{cl}(X_i) \cap X_j \neq \emptyset$.

Show that if $X_{i(1)}, \ldots, X_{i(r)}$ are the vertices of a connected component of this graph, then $X_{i(1)} \cup \cdots \cup X_{i(r)}$ is a definably connected commponent of X, and that all definably connected components of X are of this form. (This gives a useful method to construct the definably connected components of X from a partition of X into finitely many cells.)

6. Suppose $\mathcal{S}'$ is an o-minimal structure on $(R, <)$ with $\mathcal{S} \subseteq \mathcal{S}'$, and let $X \subseteq R^m$ belong to $\mathcal{S}$, so X also belongs to $\mathcal{S}'$. Show that X is definably connected in the sense of $\mathcal{S}$ if and only if X is definably connected in the sense of $\mathcal{S}'$.

7. Suppose $\mathcal{S}$ is an o-minimal structure on the ordered set $(\mathbb{R}, <)$ of real numbers. Show that for a definable set $X \subseteq \mathbb{R}^m$ the following are equivalent:

(a) X is definably connected;

(b) X is connected in the usual topological sense.

REMARK. If $\mathcal{S}$ contains addition, then (a) and (b) above are also equivalent to "X is definably path connected", see Chapter 6, (3.2), and also Chapter 6, (1.15), exercise 8.

8. With the same hypothesis as in exercise 7, show that each definable set $X \subseteq \mathbb{R}^m$ is locally connected, that is, for each $x \in X$ and each open subset U of X containing x there is a connected open subset V of X containing x and contained in U.

§3. Definable families

(3.1) Let $S \subseteq R^{m+n} = R^m \times R^n$ be definable. For each $a \in R^m$ we put

$$S_a := \{x \in R^n : (a, x) \in S\}, \text{ a subset of } R^n.$$

We view S as describing the family of sets $(S_a)_{a \in R^m}$. Such a family is called a **definable family** (of subsets of R^n, with parameter space R^m). The sets S_a are also called the **fibers** of the family.

(3.2) EXAMPLE. Let $\mathcal{R} := (\mathbb{R}, <, +, \cdot)$, and consider the formula

$$(*) \qquad ax^2 + bxy + cy^2 + dx + ey + f = 0.$$

This defines a relation $S \subseteq \mathbb{R}^6 \times \mathbb{R}^2$. For each point (a, b, c, d, e, f) in $\mathbb{R}^6$ the subset $S_{(a,b,c,d,e,f)}$ of $\mathbb{R}^2$ consists of the points (x, y) satisfying $(*)$. We know that such a set can be an ellipse, a parabola, a hyperbola, two intersecting lines, two parallel lines, a single line, a single point, the empty set, or the entire plane $\mathbb{R}^2$. In total seven different homeomorphism types occur among the sets in the family $(S_a)_{a \in \mathbb{R}^6}$.

(3.3) In Chapter 9 we prove more generally for o-minimal expansions of ordered rings that the sets of any given definable family belong to only finitely many definable homeomorphism types. (Two definable sets belong to the same definable homeomorphism type if there is a definable homeomorphism between them.) This is not true for arbitrary o-minimal structures: all intervals of any given o-minimal structure belong to a single definable family, but given any cardinal κ there are o-minimal structures having κ mutually non-homeomorphic intervals, see Chapter 1, (6.6) and (7.11). Nevertheless, we will prove here some finiteness results on definable families in arbitrary o-minimal structures that are needed in the next chapters.

(3.4) In the following $\pi : R^{m+n} \to R^m$ denotes the projection on the first m coordinates.

(3.5) PROPOSITION.

(i) *Let C be a cell in R^{m+n} and $a \in \pi(C)$. Then C_a is a cell in R^n.*

(ii) *Let $\mathcal{D}$ be a decomposition of R^{m+n} and $a \in R^m$. Then the collection*

$$\mathcal{D}_a := \{C_a : C \in \mathcal{D},\ a \in \pi(C)\}$$

is a decomposition of R^n.

PROOF. For $n = 1$ this is immediate from the definitions. Suppose the proposition holds for a certain n, and let C be a cell in $R^{m+(n+1)}$. Let $\pi_1 : R^{m+(n+1)} \to R^{m+n}$ be the obvious projection map, so that $\pi \circ \pi_1 : R^{m+(n+1)} \to R^m$ is the projection on the first m coordinates.

If $C = \Gamma(f)$, then $C_a = \Gamma(f_a)$, where $f_a : (\pi_1 C)_a \to R$ is defined by $f_a(x) = f(a, x)$. If $C = (f, g)_D$ with $D = \pi_1 C$, then $C_a = (f_a, g_a)_E$, where $E = D_a$.

In both cases it is clear that C_a is a cell in R^{n+1}. Property (ii) follows easily. □

(3.6) COROLLARY. *Let $S \subseteq R^m \times R^n$ be definable. Then there is a number $M_S \in \mathbf{N}$ such that for each $a \in R^m$ the set $S_a \subseteq R^n$ has a partition into at most M_S cells. In particular, each fiber S_a has at most M_S definably connected components.*

PROOF. Take a decomposition $\mathcal{D}$ of R^{m+n} partitioning S. Then for each a in R^m the decomposition $\mathcal{D}_a = \{C_a : C \in \mathcal{D},\ a \in \pi C\}$ of R^m consists of at most $|\mathcal{D}|$ cells and partitions S_a. So we can take $M_S = |\mathcal{D}|$. □

The following consequence is worth recording.

(3.7) COROLLARY. *Let $S \subseteq R^m \times R^n$ be definable. Then there is a natural number M_S such that for each $a \in R^m$ the set $S_a \subseteq R^n$ has at most M_S isolated points. In particular, each finite fiber S_a has cardinality at most M_S.*

(3.8) The following exercise is intended for readers familiar with elementary logic.

EXERCISE.

Let $\mathcal{R} = (R, <, \ldots)$ be an o-minimal L-structure and $\mathcal{R}' = (R', <', \ldots)$ an L-structure elementarily equivalent to $\mathcal{R}$. Show that $\mathcal{R}'$ is also o-minimal.

Notes and comments

The monotonicity theorem for o-minimal structures on the real line has an easy proof, see Van den Dries [19]. The more general monotonicity theorem (1.2) is due to Pillay and Steinhorn [49]. The cell decomposition theorem for "strongly" o-minimal structures on the real line is also in [19]. The general cell decomposition theorem (2.11) is again more difficult and was established by Knight, Pillay and Steinhorn in [35]. It is worth keeping in mind that most proofs that a particular structure is o-minimal actually give at the same time the uniform finiteness property (2.13), that is, strong o-minimality. Thus we could have simplified this chapter considerably for those readers who want to restrict their attention to o-minimal structures on the real line for which the finiteness property (2.13) is available (as is usually the case). However, in the model-theoretic literature on o-minimal structures it is important to have the results available in the present generality, sometimes even in the sharper form given in (1.9), exercise 1, in (2.19), exercise 4, and in the exercise of (3.8).

CHAPTER 4

DEFINABLE INVARIANTS: DIMENSION AND EULER CHARACTERISTIC

As before we fix an o-minimal structure $(R, <, \mathcal{S})$, and the usual terminological and notational conventions of Chapters 1 and 3 remain in force. In this chapter we establish the basic properties of the dimension and the Euler characteristic of a definable set. Dimension and Euler characteristic are definable invariants for two reasons: they are invariant under definable bijections, and they vary "definably" in a definable family. At the end of Section 1 we apply results on dimension to construct stratifications of definable sets.

§1. Dimension

(1.1) We define the **dimension** of a nonempty definable set $X \subseteq R^m$ by

$$\dim X := \max\{i_1 + \cdots + i_m : X \text{ contains an } (i_1, \ldots, i_m)\text{-cell}\}.$$

To the empty set we assign the dimension $-\infty$.

So $\dim X \in \{-\infty, 0, 1, \ldots, m\}$, and $\dim X = m$ iff X contains an open cell. Partitioning X into finitely many cells we see that $\dim X = 0$ iff X is finite and nonempty. To prove that this dimension function has the right properties we need the following.

(1.2) LEMMA. *If $A \subseteq R^m$ is an open cell and $f : A \to R^m$ an injective definable map, then $f(A)$ contains an open cell.*

PROOF. Clear for $m = 1$. Let $m > 1$ and assume inductively the lemma holds for lower values of m. Taking a decomposition of R^m that partitions $f(A)$ we have

$$f(A) = C_1 \cup \cdots \cup C_k \text{ for cells } C_i \text{ in } R^m.$$

Then

$$A = f^{-1}(C_1) \cup \cdots \cup f^{-1}(C_k),$$

so at least one of the $f^{-1}(C_i)$, say $f^{-1}(C_1)$, contains a box B, and by taking B suitably small we may assume that $f|B$ is continuous. (By cell decomposition.) We now claim that C_1 is open. If not, then by composing $f|B: B \to C_1$ with a definable homeomorphism of C_1 with a cell in R^{m-1} we obtain a definable continuous injective map $g: B \to R^{m-1}$. Write $B = B' \times (a, b)$.

Take c with $a < c < b$ and consider the map $h: B' \to R^{m-1}$ given by $h(x) = g(x, c)$. By the inductive assumption applied to h we get $h(B') \supseteq D$ for some box D in R^{m-1}. Let y be a point in D and take x in B' with $h(x) = y$.

If $c' \neq c$ is sufficiently close to c, then $g(x, c')$ will be in D, so $g(x, c') = h(x') = g(x', c)$ for some x' in B'. This contradicts the injectivity of g. □

(1.3) Proposition.

(i) *If $X \subseteq Y \subseteq R^m$ and X, Y are definable, then* $\dim X \leq \dim Y \leq m$.
(ii) *If $X \subseteq R^m$ and $Y \subseteq R^n$ are definable and there is a definable bijection between X and Y, then* $\dim X = \dim Y$.
(iii) *If $X, Y \subseteq R^m$ are definable, then* $\dim(X \cup Y) = \max\{\dim X, \dim Y\}$.

Proof. Property (i) is obvious. To prove (ii), let $f: X \to Y$ be a definable bijection and $d = \dim X$, $e = \dim Y$. It is enough to show $d \leq e$ since the reverse inequality then follows by using f^{-1}. Let A be an $(i_1, \ldots, i_m)$-cell contained in X, with $d = i_1 + \cdots + i_m$. Then $f \circ (p_A^{-1}) : p(A) \to Y$ is an injective map and $p(A)$ an open cell. Replacing X by $p(A)$, Y by $f(A)$ and f by $f \circ (p_A^{-1})$ we may as well assume that $d = m$ and that X is an open cell in R^d. Let $Y = C_1 \cup \cdots \cup C_k$ be a partition of $Y = f(X)$ into cells. Then $X = f^{-1}(C_1) \cup \cdots \cup f^{-1}(C_k)$, so by the cell decomposition theorem $f^{-1}(C_i)$ contains an open cell B, for some i. Fix such i and B.

Let $C_i = C \subseteq R^n$ be a $(j_1, \ldots, j_n)$-cell. We shall prove that $d \leq j_1 + \cdots + j_n$. (Since $j_1 + \cdots + j_n \leq e$ this will finish the proof.)

Suppose $d > j_1 + \cdots + j_n$. The composition

$$B \xrightarrow{f|B} C \xrightarrow{p_C} p(C) \subseteq R^{j_1 + \cdots + j_n}$$

is an injective map. Identifying $R^{j_1 + \cdots + j_n}$ with a non-open cell $(R^{j_1 + \cdots + j_n}) \times \{p\}$ in R^d, where $p \in R^{d - (j_1 + \cdots + j_n)}$, we obtain a contradiction with lemma (1.2).

To prove (iii), let $d = \dim(X \cup Y)$, and let A be an $(i_1, \ldots, i_m)$-cell contained in $X \cup Y$, with $d = i_1 + \cdots + i_m$. The open cell $pA \subseteq R^d$ is the union of $p_A(A \cap X)$ and $p_A(A \cap Y)$, so by the cell decomposition theorem, one of these sets, say $p_A(A \cap X)$, contains a box B in R^d. Then $p_A^{-1}(B)$ is an $(i_1, \ldots, i_m)$-cell contained in X, so that

$$\dim X \geq d \geq \dim X, \text{ i.e., } \dim X = \dim(X \cup Y). \quad \square$$

(1.4) Our definition of dimension was admittedly ad hoc, but is vindicated by its invariance under definable bijections (property (ii) of the proposition above). As a special case of this invariance, an $(i_1,\ldots,i_m)$-cell A has dimension $i_1+\cdots+i_m$. (Use the bijection p_A between A and an open cell in $R^{i_1+\cdots+i_m}$.)

The next result says among other things that the dimension of a set from a definable family depends "definably" on its parameters.

(1.5) PROPOSITION. *Let $S\subseteq R^m\times R^n$ be definable. For $d\in\{-\infty,0,1,\ldots,n\}$ put*

$$S(d):=\{a\in R^m:\ \dim S_a=d\}.$$

Then $S(d)$ is definable and the part of S above $S(d)$ has dimension given by

$$\dim\left(\bigcup_{a\in S(d)}\{a\}\times S_a\right)=\dim(S(d))+d.$$

PROOF. Let $\mathcal{D}$ be a decomposition of R^{m+n} partitioning S. Let C be a cell in $\mathcal{D}$, with projection $\pi C\subseteq R^m$. If C is an $(i_1,\ldots,i_m,\ldots,i_{m+n})$-cell, then πC is an $(i_1,\ldots,i_m)$-cell and C_a is an $(i_{m+1},\ldots,i_{m+n})$-cell for each $a\in\pi C$. (By induction on n, see Chapter 3, (3.5)(i) and its proof.) Hence,

$$(*)\qquad \dim C=\dim(\pi C)+\dim C_a,\text{ for each } a\in\pi C.$$

Consider now a cell $A\in\pi\mathcal{D}=\{\pi C:\ C\in\mathcal{D}\}$, and let $C_1,\ldots,C_k$ be the cells in $\mathcal{D}$ that are contained in S and that project onto A, i.e., $\pi C_1=\cdots=\pi C_k=A$.

For each $a\in A$ we have $S_a=(C_1)_a\cup\cdots\cup(C_k)_a$, so

$$\begin{aligned}\dim(S_a)&=\sup_{1\le i\le k}\dim(C_i)_a\\ &=\sup_{1\le i\le k}(\dim C_i-\dim A),\text{ by }(*).\end{aligned}$$

Calling this supremum d we note that $\dim S_a=d$ is independent of $a\in A$, therefore $A\subseteq S(d)$. So $S(d)$ is a union of cells in $\pi(\mathcal{D})$, hence $S(d)$ is definable. Note also that

$$\begin{aligned}d=\left(\sup_{1\le i\le k}\dim C_i\right)-\dim A&=\dim\left(\bigcup_{1\le i\le k}C_i\right)-\dim A\\ &=\dim\left(\bigcup_{a\in A}\{a\}\times S_a\right)-\dim A,\end{aligned}$$

i.e., $\dim\left(\bigcup_{a\in A}\{a\}\times S_a\right)=\dim A+d$, and taking the union over all $A\in\pi\mathcal{D}$ with $A\subseteq S(d)$ we get

$$\dim\left(\bigcup_{a\in S(d)}\{a\}\times S_a\right)=\dim(S(d))+d.\quad\square$$

(1.6) Corollary.

(i) $\dim S = \max_{0 \le d \le n}(\dim S(d) + d) \ge \dim \pi S$.

(ii) *Let $X \subseteq R^n$ be definable and $f: X \to R^m$ a definable map. Then for each $d \in \{0, \ldots, n\}$ the set $S_f(d) := \{a \in R^m : \dim f^{-1}(a) = d\}$ is definable and $\dim f^{-1}(S_f(d)) = \dim S_f(d) + d$. Moreover, $\dim X \ge \dim f(X)$.*

(iii) $\dim(A \times B) = \dim A + \dim B$, *for definable sets A and B.*

Proof. The equality in (i) is immediate from (1.5) and (1.3)(iii), and the inequality from $\pi S = \bigcup_{0 \le d \le n} S(d)$ and (1.3)(iii).

Property (ii) follows from (i) and (1.5) by letting $S = \{(f(x), x) : x \in X\}$, taking into account (1.3)(ii) and the fact that $x \mapsto (f(x), x)$ is a definable bijection between X and S. Property (iii) is a special case of (i) by letting $S = A \times B$. □

Note that in the situation of (ii) we have $\dim X = \dim f(X)$ if each fiber $f^{-1}(a)$ is finite.

We now turn to a more delicate result on dimension. Recall that for any definable set $S \subseteq R^m$ we call $\mathrm{cl}(S) - S$ the **frontier** of S, to be distinguished from the boundary $\mathrm{bd}(S) := \mathrm{cl}(S) - \mathrm{int}(S)$. Notation: $\partial S := \mathrm{cl}(S) - S$. We shall prove below that $\dim \partial S < \dim S$ if $S \ne \emptyset$. First a technical lemma.

(1.7) Lemma. *Let $m > 0$ and $A \subseteq R^m$ be definable. Then the set*

$$A' := \{x \in R : \partial(A_x) \ne (\partial A)_x\}$$

is finite.

Proof. Note that $\partial(A_x) \subseteq (\partial A)_x$ for all $x \in R$. Assume for a contradiction that A' is infinite. Then A' contains an interval I. After replacing A by $A \cap (I \times R^{m-1})$ we may assume that $I = A' = \pi_1(A)$, where $\pi_1 : R^m \to R$ is the projection on the first coordinate. For $x \in I$, put $F(x) := (\partial A)_x - \partial(A_x)$, so $F(x)$ is a nonempty subset of R^{m-1}. Hence there exists a box $B \subseteq R^{m-1}$ such that $B \cap F(x) \ne \emptyset$ and $B \cap A_x = \emptyset$. For any box $B \subseteq R^{m-1}$, put

$$I_B := \{x \in I : B \cap F(x) \ne \emptyset \text{ and } B \cap A_x = \emptyset\}.$$

Let

$$G := \{(a, b) = (a_1, \ldots, a_{m-1}, b_1, \ldots, b_{m-1}) \in R^{2(m-1)} : a_i < b_i \text{ for all } i\},$$

and let $B(a, b) := (a_1, b_1) \times \cdots \times (a_{m-1}, b_{m-1})$ be the box in R^{m-1} corresponding to a point $(a, b) \in G$. We now introduce the definable set

$$C := \bigcup_{(a,b) \in G} I_{B(a,b)} \times \{(a, b)\} \quad \text{(a subset of } R^{2m-1}\text{)}.$$

We shall establish two claims:

(1) The set $I_{B(a,b)}$ is finite for every $(a, b) \in G$.

(2) $\mathrm{int}(C_x) \ne \emptyset$ for every $x \in I$.

Claim 1 and (1.5) imply $\dim C \leq \dim G = 2(m-1)$, while claim 2 and (1.5) imply $\dim C \geq 1 + 2(m-1)$, a contradiction. Thus it only remains to prove the claims.

(1) Suppose Claim 1 is false. Then there is a box $B \subseteq R^{m-1}$ and an interval $J \subseteq I_B$. By definition of I_B we have $(J \times B) \cap A = \emptyset$, hence $(J \times B) \cap \mathrm{cl}(A) = \emptyset$ since $J \times B$ is open. But for each $x \in J$ we have $B \cap F(x) \neq \emptyset$, so in particular there exists $y \in B$ such that $(x, y) \in (J \times B) \cap \mathrm{cl}(A)$, a contradiction.

(2) Let $x \in I$. Then there exists a point $(a, b) \in C_x$. Thus $F(x) \cap B(a, b) \neq \emptyset$. Take some $y \in F(x) \cap B(a, b)$. Then every point $(a', b') \in G$ such that $y \in B(a', b') \subseteq B(a, b)$ also belongs to C_x. Thus $\mathrm{int}(C_x) \neq \emptyset$.

This finishes the proof of the lemma. □

(1.8) THEOREM. *Let $S \subseteq R^m$ be a nonempty definable set. Then*

$$\dim \partial S < \dim S.$$

In particular $\dim(\mathrm{cl}(S)) = \dim S$.

PROOF. By induction on m. The cases $m = 0$ and $m = 1$ are obvious. So let $m > 1$ and assume the result holds for lower values of m. We also assume that $\dim S > 0$, since otherwise the result holds trivially. For each $i = 1, \ldots, m$ we consider the definable bijection $\phi_i : R^m \to R^m$ defined by $\phi_i(x_1, \ldots, x_m) = (x_i, x_1, \ldots, x_{i-1}, x_{i+1}, \ldots, x_m)$. We apply the preceding lemma to $\phi_i(S)$ to get a finite set $F_i \subseteq R$ such that $\partial(\phi_i(S)_x) = (\partial\phi_i(S))_x$ for all $x \in R - F_i$. Put $H_i := \pi_i^{-1}(F_i) = R^{i-1} \times F_i \times R^{m-i}$, where $\pi_i : R^m \to R$ is the projection on the ith coordinate, and put $H := \bigcap_{i=1}^m H_i$. Then $H = F_1 \times \cdots \times F_m$ is finite. Moreover

$$\partial S \subseteq H \cup ((\partial S) - H) = H \cup \bigcup_{i=1}^{m} (\partial S) - H_i.$$

Thus by (1.3) it is enough to show that $\dim((\partial S) - H_i) < \dim S$ for each i. Fix some index i. Then

$$\phi_i((\partial S) - H_i) = \bigcup_{x \in R - F_i} \{x\} \times \partial(\phi_i(S)_x).$$

By the inductive hypothesis $\dim \partial(\phi_i(S)_x) < \dim(\phi_i(S))_x$ for all $x \in R$ for which $\phi_i(S)_x \neq \emptyset$. Thus by the last formula displayed, taking only the union over the $x \in R - F_i$ such that $\phi_i(S)_x \neq \emptyset$:

$$\dim((\partial S) - H_i) = \dim \phi_i((\partial S) - H_i) < \dim \phi_i(S) = \dim S.$$

□

(1.9) COROLLARY. *Let S and T be nonempty definable subsets of R^m with $S \subseteq T$ and* $\dim S = \dim T$. *Then S has nonempty interior* $\mathrm{int}_T(S)$ *in T and*

$$\dim(S - \mathrm{int}_T(S)) < \dim S.$$

PROOF. Let cl_T denote the closure in T. Then $S - \mathrm{int}_T(S) = S \cap \mathrm{cl}_T(T - S) = \mathrm{cl}_T(T - S) - (T - S)$, which either is empty or has by the theorem above dimension $< \dim(T - S) \leq \dim S$. □

(1.10) Corollary. *Let $S \subseteq R^m$ be definable. Then* $\dim(\mathrm{bd}(S)) < m$, *where* $\mathrm{bd}(S) := \mathrm{cl}(S) - \mathrm{int}(S)$ *is the topological boundary of the set S in R^m.*

Proof. Clear if $\dim S < m$. If $\dim S = m$ one uses

$$\mathrm{bd}(S) = \big(\mathrm{cl}(S) - S\big) \cup \big(S - \mathrm{int}(S)\big)$$

together with theorem (1.8), and corollary (1.9) with $T = R^m$. □

Digression: existence of stratifications. (This material is not used later in the book.)

(1.11) Definition. A **stratification** $\mathfrak{S}$ of a closed definable set $S \subseteq R^m$ is a partition of S into finitely many cells, called strata of $\mathfrak{S}$, such that for each stratum $A \in \mathfrak{S}$ we have: ∂A is a union of (necessarily lower-dimensional) strata of $\mathfrak{S}$.

(1.12) Each partition of a closed definable subset of R into finitely many cells is a stratification, but there are obvious examples of decompositions of R^2 that are not stratifications.

(1.13) Proposition. *Let $A \subseteq R^m$ be a closed definable set and $A_1, \ldots, A_k$ definable subsets of A. Then there is a stratification of A partitioning each of $A_1, \ldots, A_k$.*

First two easy lemmas:

(1.14) Lemma. *Let C, D be cells in R^m, where C is an $(i_1, \ldots, i_m)$-cell and $D \subseteq C$. Then the following are equivalent:*

(i) *D is an $(i_1, \ldots, i_m)$-cell;*
(ii) $\dim C = \dim D$;
(iii) *D is open in C.*

Proof. A straightforward induction on m, just using the definition of cell. □

(1.15) Lemma. *If $A \subseteq R^m$ is definable and $C \subseteq A$ is a cell with* $\dim C = \dim A$, *then there are cells $D_1, \ldots, D_k \subseteq C$ that are open in A such that*

$$\dim\big(C - (D_1 \cup \cdots \cup D_k)\big) < \dim C.$$

Proof. Write $\mathrm{int}_A C = D_1 \cup \cdots \cup D_k \cup D_{k+1} \cup \cdots \cup D_l$, where $D_1, \ldots, D_k$ are cells of the same dimension as C, and $D_{k+1}, \ldots, D_l$ are cells of lower dimension. Then by lemma (1.14) each D_i with $1 \leq i \leq k$ is open in C, hence open in $\mathrm{int}_A C$, hence open in A. Moreover, $C - (D_1 \cup \cdots \cup D_k) \subseteq (C - \mathrm{int}_A C) \cup D_{k+1} \cup \cdots \cup D_l$ which has by (1.9) dimension $< \dim C$. □

(1.16) PROOF OF (1.13). By induction on $\dim A$. If $\dim A = 0$ then A is finite and the desired result is obvious. Suppose $\dim A > 0$ and that the proposition holds for definable sets of lower dimension. Take a finite partition $\mathcal{P}$ of A into cells that also partitions each A_i. Let $\mathcal{P}_0$ be the collection of cells in $\mathcal{P}$ of dimension $\dim A$. After refining $\mathcal{P}$ we may assume by lemma (1.15) that the cells of $\mathcal{P}_0$ are open in A. Then $A - \bigcup \mathcal{P}_0$ is a closed set of dimension $< \dim A$ containing ∂D for each $D \in \mathcal{P}_0$ and E for each $E \in \mathcal{P} - \mathcal{P}_0$, so by the inductive hypothesis there is a stratification $\mathfrak{S}^*$ of $A - \bigcup \mathcal{P}_0$ that refines $\mathcal{P} - \mathcal{P}_0$ and partitions the sets ∂D for all $D \in \mathcal{P}_0$. Now set $\mathfrak{S} := \mathcal{P}_0 \cup \mathfrak{S}^*$. We claim that $\mathfrak{S}$ satisfies the requirements. By construction $\mathfrak{S}$ is a stratification of A, and because $\mathfrak{S}^*$ partitions each cell of $\mathcal{P} - \mathcal{P}_0$ and $\mathcal{P}$ partitions each A_i, $\mathfrak{S}$ partitions each A_i. □

(1.17) EXERCISES.

1. Let $A \subseteq R^m$ be definable and $0 \leq d \leq m$. Show that $\dim A \geq d$ if and only if there is a d-tuple $i = \big(i(1), \ldots, i(d)\big)$ with $1 \leq i(1) < \cdots < i(d) \leq m$ such that the projection map $p_i : R^m \to R^d$ given by $p_i(x_1, \ldots, x_m) = \big(x_{i(1)}, \ldots, x_{i(d)}\big)$ has the property that $p_i(A)$ has nonempty interior in R^d.

2. Let $A \subseteq R^m$ be a definable set and $a \in R^m$. Show there is a number $d \in \{-\infty, 0, \ldots, \dim A\}$ such that $\dim(U \cap A) = d$ for all sufficiently small definable neighborhoods U of a in R^m, that is, for all definable neighborhoods of a in R^m that are contained in some fixed definable neighborhood of a in R^m.

The number d defined by this property is called the **local dimension of A at a**, notation $\dim_a(A)$. Note that $\dim_a(A) = -\infty$ iff $a \notin \mathrm{cl}(A)$.

3. Show that if A is a d-dimensional cell, then $\dim_a(A) = d$ for all $a \in \mathrm{cl}(A)$.

4. Let $A \subseteq R^m$ be a definable set and $d \in \{0, \ldots, \dim A\}$. Show that the set $\{a \in R^m : \dim_a(A) \geq d\}$ is a definable closed subset of $\mathrm{cl}(A)$. Show also that if $A \neq \emptyset$, then $\dim\big(\{a \in \mathrm{cl}(A) : \dim_a(A) < d\}\big) < d$.

§2. Euler characteristic

(2.1) Besides "dimension" there is another, more subtle, definable invariant, the Euler characteristic. As a simple example, observe that a finite partition of an interval into cells consists necessarily of k points and $k+1$ intervals, for some $k \geq 0$, and that the quantity $k - (k+1) = -1$ is independent of the partition considered.

More generally, we assign to each cell C of dimension d the integer

$$E(C) := (-1)^d,$$

and given a finite partition $\mathcal{P}$ of a definable set $S \subseteq R^m$ into cells we put

$$E_{\mathcal{P}}(S) := \sum_{C \in \mathcal{P}} E(C) = k_0 - k_1 + \cdots + (-1)^d k_d + \cdots + (-1)^m k_m,$$

where k_d is the number of d-dimensional cells in $\mathcal{P}$.

We now have the following basic result.

(2.2) PROPOSITION. *If $\mathcal{P}'$ is a second finite partition of S into cells, then we have*

$$E_{\mathcal{P}}(S) = E_{\mathcal{P}'}(S).$$

(2.3) We shall prove this below, see (2.8). Accepting the proposition for the moment we may define the **Euler characteristic** $E(S)$ of S to be the common value of $E_{\mathcal{P}}(S)$ for finite partitions $\mathcal{P}$ of S into cells. The second basic fact is the invariance of the Euler characteristic under definable bijections:

(2.4) PROPOSITION. *If $f: S \to R^n$ is an injective definable map, then*

$$E(S) = E(f(S)).$$

(2.5) This will be established in (2.12) below. Before starting the proofs of these propositions we single out certain finite partitions of cells and call these **decompositions**, generalizing the notion of decomposition of R^m from Chapter 3, (2.10). The definition is by induction:

(1) All finite partitions of a cell $C \subseteq R = R^1$ into cells are decompositions of C;

(2) For $m \geq 1$ a decomposition of a cell $C \subseteq R^{m+1}$ is a finite partition $\mathcal{D}$ of C into cells such that $\pi(\mathcal{D}) := \{\pi D : D \in \mathcal{D}\}$ is a decomposition of the cell $\pi C \subseteq R^m$.

Note: (2) also holds for $m = 0$ by considering the unique partition of the trivial cell R^0 as a decomposition. The following fact is mentioned for the sake of completeness and will not be used: the decompositions of a cell $C \subseteq R^m$ are exactly the restrictions to C of the decompositions of R^m that partition C. (Exercise)

(2.6) LEMMA. *If $\mathcal{D}$ is a decomposition of a cell C, then*

$$E_{\mathcal{D}}(C) = E(C) \; (= (-1)^{\dim C}).$$

PROOF. By induction. Clear if $C \subseteq R = R^1$. Let $C \subseteq R^{m+1}$, $m \geq 1$, and assume inductively

$$(*) \qquad E_{\pi(\mathcal{D})}(\pi C) = E(\pi C).$$

We also assume that C is an $(i_1, \ldots, i_m, 1)$-cell; the case that C is an $(i_1, \ldots, i_m, 0)$-cell is similar and left to the reader.

Let $B \in \pi(\mathcal{D})$, so the list of distinct cells of $\mathcal{D}$ that map onto B under π is

$$\Gamma(f_1), \ldots, \Gamma(f_t), (f_0, f_1), \ldots, (f_t, f_{t+1}),$$

for certain continuous definable functions f_i on B. Let $\dim B = d$. The contribution of this list of cells to $E_{\mathcal{D}}(C)$ equals $t \cdot (-1)^d + (t+1) \cdot (-1)^{d+1} = (-1)^{d+1} = -E(B)$. Summing over the various $B \in \pi(\mathcal{D})$ we get, using $(*)$,

$$E_{\mathcal{D}}(C) = -\sum_{B \in \pi(\mathcal{D})} E(B) = -E_{\pi(\mathcal{D})}(\pi C) = -E(\pi C) = E(C). \quad \square$$

(2.7) LEMMA. *Every two finite partitions $\mathcal{P}(1)$ and $\mathcal{P}(2)$ of a definable set $S \subseteq R^m$ into cells have a common refinement to a finite partition $\mathcal{P}$ of S into cells such that for each cell $C \in \mathcal{P}(1) \cup \mathcal{P}(2)$ the restriction $\mathcal{P}|C$ is a decomposition of C.*

PROOF. Take a decomposition of R^m that partitions each cell of $\mathcal{P}(1) \cup \mathcal{P}(2)$. Restricting this decomposition to S gives a partition $\mathcal{P}$ as required. $\square$

(2.8) PROOF OF PROPOSITION (2.2). Let $\mathcal{P}(1)$ and $\mathcal{P}(2)$ be finite partitions of the definable set $S \subseteq R^m$ into cells. We have to show that $E_{\mathcal{P}(1)}(S) = E_{\mathcal{P}(2)}(S)$. Take a common refinement $\mathcal{P}$ of $\mathcal{P}(1)$ and $\mathcal{P}(2)$ with the property of the lemma above. Then

$$\begin{aligned} E_{\mathcal{P}}(S) &= \sum_{C \in \mathcal{P}(1)} E_{\mathcal{P}|C}(C) \\ &= \sum_{C \in \mathcal{P}(1)} E(C) \quad \text{by lemma (2.6)} \\ &= E_{\mathcal{P}(1)}(S), \end{aligned}$$

and in the same way we get $E_{\mathcal{P}}(S) = E_{\mathcal{P}(2)}(S)$. $\square$

(2.9) As we noted in (2.3) we may now speak of the Euler characteristic $E(S)$ of a definable set S without specifying a partition of S into cells. Clearly, for definable sets $S_1, S_2 \subseteq R^m$ we have

$$E(S_1 \cup S_2) = E(S_1) + E(S_2) \text{ if } S_1 \text{ and } S_2 \text{ are disjoint;}$$

in general

$$E(S_1 \cup S_2) = E(S_1) + E(S_2) - E(S_1 \cap S_2),$$

the second formula following from the first by representing $S_1 \cup S_2$ as the disjoint union of $S_1 - (S_1 \cap S_2)$, $S_1 \cap S_2$ and $S_2 - (S_1 \cap S_2)$.

Next we show that the Euler characteristic varies "definably" in a definable family of sets.

(2.10) PROPOSITION. *Let $S \subseteq R^{m+n}$ be definable. Then $E(S_a)$ takes only finitely many values as a runs through the parameter space R^m, and for each integer e the set $\{a \in R^m : E(S_a) = e\}$ is definable. More precisely, let $\mathcal{D}$ be a decomposition of R^{m+n} partitioning S and let $\pi : R^{m+n} \to R^m$ be the projection on the first m coordinates. Given a cell $A \in \pi(\mathcal{D})$ there is a constant $e_A \in \mathbf{Z}$ such that $E(S_a) = e_A$ for all $a \in A$; also*

$$E(\pi^{-1}(A) \cap S) = E(A)e_A.$$

PROOF. Suppose $A \in \pi(\mathcal{D})$ is an $(i(1), \ldots, i(m))$-cell. Let $C \in \mathcal{D}$ be a cell contained in S with $\pi C = A$, so C is an $(i(1), \ldots, i(m), i(m+1), \ldots, i(m+n))$-cell for certain $i(m+1), \ldots, i(m+n)$. For each $a \in A$ the fiber C_a is an $(i(m+1), \ldots, i(m+n))$-cell, so that $E(C) = E(A) \cdot E(C_a)$ for all $a \in A$. Since $\pi^{-1}(A) \cap S$ is the union of cells $C \in \mathcal{D}$, and S_a for $a \in A$ is the union of the corresponding cells C_a, we obtain the constancy of $E(S_a)$ as well as the formula for $E(\pi^{-1}(A) \cap S)$. $\square$

(2.11) COROLLARY. *Let $S \subseteq R^{m+n}$ be definable and suppose all nonempty fibers S_a $(a \in R^m)$ have the same Euler characteristic e and let $\pi : R^{m+n} \to R^m$ be as above. Then*

$$E(S) = E(\pi S) \cdot e,$$

in particular $E(A \times B) = E(A) \cdot E(B)$ for definable $A \subseteq R^m$ and $B \subseteq R^n$.

(2.12) We can now start the proof of proposition (2.4). When finished we will have removed the blemish that our Euler characteristic seems to depend on the notion of "cell" which is not invariant under coordinate permutations.

Let $S \subseteq R^m$ be definable and $f : S \to R^n$ an injective definable map. We have to show that $E(S) = E(f(S))$. (Note: we do not require continuity of f.)

By applying (2.11) to $\Gamma(f)$ we see that $E(S) = E(\Gamma(f))$. Let $\Gamma'(f) := \{(f(x), x) : x \in S\}$ be the "reversed" graph of f. Again by (2.11) we have $E(f(S)) = E(\Gamma'(f))$. If we knew that $E(\Gamma(f)) = E(\Gamma'(f))$ then we could conclude that $E(S) = E(f(S))$. So we have reduced to proving the following:

Let σ be a permutation of $\{1, \ldots, m\}$; setting $x\sigma := (x_{\sigma(1)}, \ldots, x_{\sigma(m)})$ for $x = (x_1, \ldots, x_m) \in R^m$, and $A\sigma := \{x\sigma : x \in A\}$ for $A \subseteq R^m$, we have $E(A) = E(A\sigma)$ for definable $A \subseteq R^m$.

Since the symmetric group on $\{1, \ldots, m\}$ is generated by the transpositions $(i, i+1)$ with $1 \leq i < m$ and A is a finite disjoint union of cells it suffices to prove this when σ is a transposition $(i, i+1)$ and A is a cell. For such σ and A this will follow if we can show that A is a finite disjoint union of subcells $A = A_1 \cup \cdots \cup A_k$ such that $(A_1)\sigma, \ldots, (A_k)\sigma$ are also cells, because then $\dim A_j = \dim A_j\sigma$ gives $E(A_j) = E(A_j\sigma)$, so that $E(A) = \sum E(A_j) = \sum E(A_j\sigma) = E(A\sigma)$. So we are done once the following proposition is established.

(2.13) PROPOSITION. *Let $C \subseteq R^m$ be a cell and $\sigma = (i, i+1)$ a transposition, $1 \leq i < m$. Then C can be partitioned into cells $C_1, \ldots, C_k$ such that $C_1\sigma, \ldots, C_k\sigma$ are also cells.*

PROOF. By induction on m and $\dim C$. Keep in mind that $\sigma = \sigma^{-1}$.

Easy case: $i < m-1$. Let us denote the restriction of σ to $\{1, \ldots, m-1\}$ also by σ, and write elements of R^m as (x, y) with $x = (x_1, \ldots, x_{m-1}) \in R^{m-1}$.

Subcase 1. $C = (\alpha, \beta)_B$. Then

$$\begin{aligned}(x,y)\in C\sigma &\Leftrightarrow (x\sigma, y) \in C \\ &\Leftrightarrow x\sigma \in B \text{ and } \alpha(x\sigma) < y < \beta(x\sigma) \\ &\Leftrightarrow x \in B\sigma \text{ and } \alpha\sigma(x) < y < \beta\sigma(x),\end{aligned}$$

where $\alpha\sigma : B\sigma \to R_\infty$ is defined by $\alpha\sigma(x) = \alpha(x\sigma)$, and similarly $\beta\sigma$.

Assuming inductively that the cell B is the disjoint union of cells $B_1, \ldots, B_k$ such that $B_1\sigma, \ldots, B_k\sigma$ are also cells we see that $C\sigma$ is the disjoint union of the k cells

$$(\alpha\sigma|B_1\sigma, \beta\sigma|B_1\sigma), \text{ which equals } (\alpha|B_1, \beta|B_1)\sigma,$$
$$\vdots$$
$$(\alpha\sigma|B_k\sigma, \beta\sigma|B_k\sigma), \text{ which equals } (\alpha|B_k, \beta|B_k)\sigma.$$

Subcase 2. $C = \Gamma(\beta)$ with $\beta : B \to R$. The argument is similar.

Hard case: $i = m-1$. Let C be a $(j_1, \ldots, j_m)$-cell and $B = \pi C$, $A = \pi' B$ where $\pi : R^m \to R^{m-1}$ and $\pi' : R^{m-1} \to R^{m-2}$ are the obvious projection maps. So B is a $(j_1, \ldots, j_{m-1})$-cell. We distinguish four subcases according to the values of j_{m-1} and j_m. We write elements of R^m as (x, y, z) with $x \in R^{m-2}$.

Subcase 3. $j_{m-1} = j_m = 0$. Then one checks easily that

$$C = \{(x, f(x), g(x)) : x \in A\}$$

for continuous definable functions $f, g : A \to R$. Then

$$C\sigma = \{(x, g(x), f(x)) : x \in A\}$$

is clearly also a cell.

Subcase 4. $j_{m-1} = 0, j_m = 1$. Then $B = \Gamma(\alpha)$ for continuous definable $\alpha : A \to R$, and $C = (f, g)$ for continuous definable $f, g : B \to R_\infty$. Hence

$$\begin{aligned}(x,y,z)\in C &\Leftrightarrow x \in A \text{ and } y = \alpha(x) \text{ and } f(x,y) < z < g(x,y) \\ &\Leftrightarrow x \in A \text{ and } f(x, \alpha(x)) < z < g(x, \alpha(x)) \text{ and } y = \alpha(x).\end{aligned}$$

Define $f', g' : A \to R_\infty$ by $f'(x) := f(x, \alpha(x))$ and $g'(x) := g(x, \alpha(x))$, and define $\alpha' : (f', g') \to R$ by $\alpha'(x, z) := \alpha(x)$. Then one verifies immediately that $C\sigma = \Gamma(\alpha')$, so $C\sigma$ is a cell.

For the remaining cases we need some further notation and terminology: given a definable continuous function $f : (a, b) \to R$ on an interval we call $c \in (a, b)$ a **critical point of** f if there are a "left" interval $(l, c) \subseteq (a, b)$ and a "right" interval $(c, r) \subseteq (a, b)$ such that

either f is strictly increasing on (l, c) and f is decreasing on (c, r),
or f is constant on (l, c) and f is strictly monotone on (c, r),
or f is strictly decreasing on (l, c) and f is increasing on (c, r).

By the monotonicity theorem f has only finitely many critical points. Let those critical points be $a_1 < \cdots < a_k$ and put $a_0 = a$, $a_{k+1} = b$. Then we associate to f the tuples $c(f) := (a_1, \ldots, a_k) \in R^k$ and $\epsilon(f) := (\epsilon_0, \ldots, \epsilon_k)$ with $\epsilon_j \in \{-1, 0, 1\}$ and

$$\begin{aligned} \epsilon_j &= -1 \text{ if } f \text{ is strictly decreasing on } (a_j, a_{j+1}), \\ \epsilon_j &= 0 \quad \text{if } f \text{ is constant on } (a_j, a_{j+1}), \\ \epsilon_j &= +1 \text{ if } f \text{ is strictly increasing on } (a_j, a_{j+1}). \end{aligned}$$

Subcase 5. $j_{m-1} = 1$, $j_m = 0$. Then $B = (\alpha, \beta)_A$ and $C = \Gamma(f)$ for a continuous definable function $f : B \to R$. The argument that follows is easy to visualize for $m = 2$ in which case $A = R^0$. The case for arbitrary m is essentially a "parametric" version of the argument for the case $m = 2$, $x \in A$ being the parameter. For each $x \in A$ we put $c(x) := c(f(x, -))$ and $\epsilon(x) := \epsilon(f(x, -))$, where of course $f(x, -) : (\alpha(x), \beta(x)) \to R$.

By partitioning A into finitely many subcells we may assume without loss of generality that $c(x)$ is of constant length k for $x \in A$, that $c(x) := (\alpha_1(x), \ldots, \alpha_k(x))$ is continuous as an R^k-valued function of $x \in A$ and that $\epsilon(x)$ is constant on A. Then B is the disjoint union of the cells (α_j, α_{j+1}) $(0 \le j \le k$, with $\alpha_0 = \alpha$ and $\alpha_{k+1} = \beta)$ and the cells $\Gamma(\alpha_j)$ $(1 \le j \le k)$, and again there is no loss of generality in assuming B is actually one of those cells; when $B = \Gamma(\alpha_j)$ we are back in subcase 3, so we may assume $B = (\alpha_j, \alpha_{j+1})$ and we rename $\alpha := \alpha_j$ and $\beta := \alpha_{j+1}$. So either $f(x, -)$ is strictly decreasing on $(\alpha(x), \beta(x))$ for all $x \in A$, or $f(x, -)$ is constant on $(\alpha(x), \beta(x))$ for all $x \in A$, or $f(x, -)$ is strictly increasing on $(\alpha(x), \beta(x))$ for all $x \in A$.

Assume that $f(x, -)$ is strictly increasing on $(\alpha(x), \beta(x))$ for all $x \in A$. (The other two possibilities are handled similarly.) Define functions $i, s : A \to R_\infty$ by

$$\begin{aligned} i(x) &:= \inf\{f(x, y) : y \in (\alpha(x), \beta(x))\}, \\ s(x) &:= \sup\{f(x, y) : y \in (\alpha(x), \beta(x))\}. \end{aligned}$$

After partitioning A further into cells we may assume i and s are continuous on A, and either R-valued, or identically $-\infty$, or identically $+\infty$. In any case $i < s$ on A.

For each $x \in A$ and $z \in (i(x), s(x))$ let $g(x,z)$ be the unique $y \in (\alpha(x), \beta(x))$ such that $f(x,y) = z$, so $f(x,y) = z \Leftrightarrow g(x,z) = y$. Then g is a definable function on $(i,s)_A$. We claim that g is continuous, from which the desired result follows since $C\sigma = \Gamma(g)$. Let $f(x,y) = z$, so $g(x,z) = y$, and let $y_1, y_2 \in R$ be such that $\alpha(x) < y_1 < y < y_2 < \beta(x)$. We have to find a neighborhood of the point (x,z) in (i,s) that is mapped into the interval (y_1, y_2) by the function g. Choose elements $z_1, z_2 \in R$ with $f(x,y_1) < z_1 < z < z_2 < f(x,y_2)$, and a neighborhood U of x in A so small that if $x' \in U$, then $\alpha(x') < y_1 < y_2 < \beta(x')$, and $f(x',y_1) < z_1 < z_2 < f(x',y_2)$. Then $U \times (z_1,z_2)$ is the desired neighborhood of (x,z): Let $(x',z') \in U \times (z_1,z_2)$. Since $f(x',-)$ is strictly increasing there is $y' \in (y_1,y_2)$ with $f(x',y') = z'$; hence $g(x',z') = y' \in (y_1,y_2)$.

Subcase 6. $j_{m-1} = j_m = 1$. Then $C = (f_1,f_2)_B$, $B = (\alpha,\beta)_A$. There are four sub-subcases depending on whether or not $f_1 = -\infty$, and whether or not $f_2 = +\infty$. To shorten the exposition we do only one of these, leaving the other three to the reader. (The ideas involved are similar.) So we shall assume that f_1 is R-valued and $f_2 = +\infty$, and we put $f := f_1$, so $C = (f, +\infty)_B$. As in subcase 5 we may reduce to the situation that $f(x,-)$ is either strictly decreasing for each $x \in A$, or constant for each $x \in A$, or strictly increasing for each $x \in A$.

Assuming the last and defining $i, s: A \to R_\infty$ as in subcase 5, we shall further assume (as in subcase 5) that i and s are continous and R-valued. (By partitioning A we may reduce to this situation, plus a few other cases that are handled similarly.) Then we define $g:(i,s) \to R$ as in subcase 5: $g(x,z) = y \Leftrightarrow f(x,y) = z$, so that g is continuous. Now define $s': B \to R$ by setting $s'(x,y) := s(x)$. Then C is the disjoint union of the cells $(f,s')_B$, $\Gamma(s')$ and $(s',+\infty)_B$ and we have $(f,s')_B\sigma = (\alpha', g)_{(i,s)}$ where $\alpha': (i,s) \to R$ is given by $\alpha'(x,z) = \alpha(x)$; $\Gamma(s')\sigma$ and $(s',+\infty)_B\sigma$ are similarly seen to be cells. This completes subcase 6, and the proofs of propositions (2.13) and (2.4). $\square$

(2.14) REMARKS. Given definable sets $A \subseteq R^m$ and $B \subseteq R^n$, we have shown that if there is a definable bijection between them, then $\dim A = \dim B$ and $E(A) = E(B)$. In Chapter 8 we shall prove the converse for o-minimal expansions of real closed fields. Without such an extra assumption one cannot expect a converse, since intervals are all of dimension 1 and Euler characteristic -1, but there are divisible ordered abelian groups in which some intervals are countable and other intervals are uncountable, see Chapter 1, (7.11).

Note that we cannot have a definable bijection between a definable set $A \subseteq R^m$ and a subset $A - \{a\}$, $a \in A$, since their Euler characteristics differ. This fact gives a negative answer to the following model-theoretic question:

Are there a definable set $A \subseteq R^n$ and a definable map $\alpha: A \to A$ such that $(A, \alpha) \equiv (\mathbf{N}, S)$, where S is the successor function on the set $\mathbf{N}$ of natural numbers?

(2.15) DIGRESSION: EULER CHARACTERISTIC OF CONSTRUCTIBLE SETS. (The material in this subsection is not used further on.)

Let K be an algebraically closed field. Recall that a constructible set in K^m is a finite union of sets of the form

$$\{x \in K^m : f_1(x) = \cdots = f_r(x) = 0,\ g_1(x) \neq 0, \ldots, g_s(x) \neq 0\},$$

where $f_i, g_j \in K[X_1, \ldots, X_m]$. The constructible sets in K^m for $m = 0, 1, 2, \ldots$ form a structure on K, in fact they are exactly the sets definable in the field K using constants, by Chevalley's constructibility theorem, see Chapter 1, (2.1).

Suppose now that K is also of characteristic 0.

Then $K = R(\mathrm{i})$ for some real closed subfield R and $\mathrm{i}^2 = -1$. (See for example Lang [37] or the next volume.) Then we may identify K with R^2 by letting $a + b\mathrm{i}$ correspond to $(a, b) \in R^2$. Then K^m gets identified with R^{2m}, a constructible set $S \subseteq K^m$ becomes a semialgebraic set $S \subseteq R^{2m}$, and a constructible map $f : S \to K^n$ becomes a semialgebraic map $f : S \to R^{2n}$; as a semialgebraic set, S has an Euler characteristic $E(S)$, with $E(S) = E\big(f(S)\big)$ if f is injective.

Does $E(S)$ depend on the choice of the real closed subfield R?

In fact, $E(S)$ is independent of the choice of R, and this is because the integer-valued function E (for any choice of R) satisfies the following rules:

(1) $E(A \cup B) = E(A) + E(B)$ for disjoint constructible $A, B \subseteq K^m$,
(2) $E(B) = e \cdot E(A)$ if $f : B \to A$ is a constructible map between constructible sets $A \subseteq K^m$ and $B \subseteq K^n$, and $e \in \mathbf{N}$ is such that $|f^{-1}(a)| = e$ for all $a \in A$,
(3) $E(K^m) = 1$ for all m.

Using known properties of constructible sets and maps one easily checks that there is at most one integer-valued function E on the class of constructible sets satisfying these three rules. Hence $E(A)$ does not depend on R.

We mention in passing that for constructible $S \subseteq K^m$ the dimension $\dim S$ of S considered as a semialgebraic set in R^{2m} is twice the dimension of the Zariski closure of S in K^m, the last dimension being taken in the sense of algebraic varieties. This shows that $\dim S$ does not depend on the choice of R either.

Notes and comments

Much of the dimension theory in Section 1 has also been developed from a more model-theoretic viewpoint by Pillay [48], using an analogue of "transcendence degree" for o-minimal structures.

Section 2 on Euler characteristic was inspired by a discussion with J. Denef, who showed me that the Euler characteristic of constructible sets over $\mathbb{C}$ is invariant under constructible bijections, and by a discussion with S. Schanuel, who told me similar results on semilinear sets and semilinear bijections; see also [51]. This suggested the possibility of a definably invariant Euler characteristic for definable sets in any o-minimal structure. Digression (2.15) relates to Denef's remarks.

Strzebonski [59] uses the "o-minimal Euler characteristic" to obtain an analogue of Sylow theory for groups definable in o-minimal structures.

Khovanskii called my attention to the following papers where Euler characteristic is viewed as a finitely additive measure in a topological setting, and integration with respect to Euler characteristic is developed and applied:

O. Ya. Viro, *Some integral calculus based on Euler characteristic*, Topology and Geometry—Rohlin Seminar, Lecture Notes in Math., vol. 1346, Springer-Verlag, Berlin, 1989, pp. 127–138.

A.V. Pukhlikov and A.G. Khovanskii, *Finitely additive measures on virtual polytopes*, St. Petersburg Math. J. **4** (1993), 337–356.

(This information came too late to include in the references at the end of the book or influence the treatment in Section 2.)

CHAPTER 5

THE VAPNIK-CHERVONENKIS PROPERTY IN O-MINIMAL STRUCTURES

Introduction

The main result in this chapter is as follows, see (3.15) below.

THEOREM. *Let* $(R, <, \mathcal{S})$ *be an o-minimal structure and* $S \subseteq R^{p+q}$ *a definable set. Then there is a positive integer* $d = d(S)$ *such that for all sufficiently large* $n \in \mathbf{N}$ *each n-element set* $F \subseteq R^q$ *has at most* n^d *subsets of the form* $S_x \cap F$ *with* $x \in R^p$.

This exhibits in a purely combinatorial way that the variation among the sets S_x is highly restricted as x ranges over the parameter space R^p: the total number of subsets of an n-element set is 2^n, and this grows much faster with n than the polynomial function n^d. The key step (taken in Section 3) is to show that a model-theoretic structure $\mathcal{R} = (R, \dots)$ has this combinatorial property, if it has the property for $q = 1$ (Shelah). The case $q = 1$ is easily checked for o-minimal structures, see (2.7) and (2.11). We also derive some other results of a rather general nature.

The combinatorial fact expressed by our theorem has an interpretation in terms of probability; see the notes at the end of the chapter. The material in this chapter will not be used later on, but it would be a pity to omit it. Moreover, I believe there is ample room for further developments along the lines of this chapter.

§1. A combinatorial dichotomy

(1.1) Let $\mathcal{C}$ be a collection of subsets of an infinite set X. Given $F \subseteq X$ we put

$$\mathcal{C} \cap F := \{C \cap F : C \in \mathcal{C}\},$$

the set of intersections of sets in $\mathcal{C}$ with F. If $A \subseteq F$ is of the form $A = C \cap F$ for some $C \in \mathcal{C}$ we also say that A **is cut out from** F **by a set in** $\mathcal{C}$. So $|\mathcal{C} \cap F| \leq 2^{|F|}$, where $|S|$ denotes the cardinality of a set S. Now we define a function $f_{\mathcal{C}} : \mathbf{N} \to \mathbf{N}$ that in some sense measures the complexity of the collection $\mathcal{C}$:

$$f_{\mathcal{C}}(n) := \max\{|\mathcal{C} \cap F| : F \text{ is an } n\text{-element subset of } X\}.$$

So $0 \leq f_{\mathcal{C}}(n) \leq 2^n$ for all n. We have the following, perhaps surprising, dichotomy.

(1.2) THEOREM. *Either $f_{\mathcal{C}}(n) = 2^n$ for all n, or else there is $d \in \mathbf{N}$ such that $f_{\mathcal{C}}(n) \leq n^d$ for all sufficiently large n.*

The first possibility means that for each n there is an n-element set $F \subseteq X$ all of whose subsets are cut out from F by sets in $\mathcal{C}$, while in the second case for all sufficiently large finite sets $F \subseteq X$ relatively few subsets of F are cut out from F by sets in $\mathcal{C}$. This theorem will be derived from the following combinatorial lemma:

(1.3) LEMMA. *Let F be a finite set of size n and $\mathcal{D} \subseteq \mathcal{P}(F)$, a collection of subsets of F, and suppose that*

$$|\mathcal{D}| > \sum_{i<d} \binom{n}{i},$$

where $0 \leq d \leq n$. Then F has a subset E such that $|E| = d$ and $\mathcal{D} \cap E = \mathcal{P}(E)$.

(1.4) REMARKS.

1. The hypothesis is sharp since the collection of subsets of F of size $< d$ has cardinality *equal* to the indicated sum of binomial coefficients, and this particular collection violates the conclusion of the lemma.

2. Let us write the indicated sum of binomial coefficients as $p_d(n)$. Clearly there is a unique polynomial $p_d(X) \in \mathbb{Q}[X]$ (of degree $d-1$ if $d \geq 1$, $p_0(X) = 0$), whose value at n is $p_d(n)$ for $n \geq d$. Note that then $p_d(x)$ is defined for arbitrary real x, and the "Pascal triangle equality" gives

$$p_{d-1}(x-1) + p_d(x-1) = p_d(x) \quad (d \geq 1).$$

(1.5) PROOF OF THE COMBINATORIAL LEMMA. By induction on n. The desired result trivially holds for $d = 0$ and $d = n$, so let $0 < d < n$. Pick a point $x \in F$ and let $F' := F - \{x\}$. Also put $D' := D - \{x\}$ for $D \in \mathcal{D}$, and let $\mathcal{D}' := \{D' : D \in \mathcal{D}\}$. Note that under the map $D \mapsto D' : \mathcal{D} \to \mathcal{D}'$ a set $S \in \mathcal{D}'$ has either exactly one preimage or exactly two preimages; in the latter case these two preimages are S and $S \cup \{x\}$. So $\mathcal{D}' = \mathcal{D}_1 \cup \mathcal{D}_2$ (a disjoint union) where $\mathcal{D}_1$ contains those $S \in \mathcal{D}'$ having one preimage in $\mathcal{D}$ and $\mathcal{D}_2$ those with two preimages. If $|\mathcal{D}'| > p_d(n-1)$ then by the inductive assumption applied to F' and $\mathcal{D}'$ there is a subset E of F' of size d with $\mathcal{D}' \cap E = \mathcal{P}(E)$, hence $\mathcal{D} \cap E = \mathcal{P}(E)$ and we are done.

So assume $|\mathcal{D}'| \leq p_d(n-1)$. But

$$\begin{aligned} |\mathcal{D}| &= |\mathcal{D}_1| + 2|\mathcal{D}_2| = (|\mathcal{D}_1| + |\mathcal{D}_2|) + |\mathcal{D}_2| \\ &= |\mathcal{D}'| + |\mathcal{D}_2| > p_d(n) = p_d(n-1) + p_{d-1}(n-1), \end{aligned}$$

hence $|\mathcal{D}_2| > p_{d-1}(n-1)$, so again by the inductive assumption applied to F' and $\mathcal{D}_2$ there is a set $E \subseteq F'$ of size $d-1$ such that $\mathcal{D}_2 \cap E = \mathcal{P}(E)$. Since for each set $S \in \mathcal{D}_2$ we have $S \in \mathcal{D}$ and $S \cup \{x\} \in \mathcal{D}$ this gives $\mathcal{D} \cap (E \cup \{x\}) = \mathcal{P}(E \cup \{x\})$. □

We now prove theorem (1.2) in the following more precise form.

(1.6) THEOREM. *Suppose d is a nonnegative integer with $f_{\mathcal{C}}(d) < 2^d$. Then $f_{\mathcal{C}}(n) \le p_d(n)$ for all n.*

PROOF. If $n < d$ then $p_d(n) = 2^n$ and the desired inequality holds trivially. So let $n \ge d$, and let F be an n-element subset of X. If $|\mathcal{C} \cap F| > p_d(n)$, then the lemma would give us a set $E \subseteq F$ of size d such that $|\mathcal{C} \cap E| = 2^d$, contradicting the assumption $f_{\mathcal{C}}(d) < 2^d$. Hence $|\mathcal{C} \cap F| \le p_d(n)$. Since F was arbitrary, the desired result follows. □

REMARK. The inequalities in the theorem are sharp, since the collection $\mathcal{C}$ of subsets of X of size $< d$ satisfies $f_{\mathcal{C}}(d) < 2^d$, $f_{\mathcal{C}}(e) = 2^e$ for $e < d$, and $f_{\mathcal{C}}(n) = p_d(n)$ for all n.

§2. Vapnik-Chervonenkis classes and dependence

(2.1) Given an infinite set X, a collection $\mathcal{C} \subseteq \mathcal{P}(X)$ is said to be a **Vapnik-Chervonenkis class** (or VC-class) if $f_{\mathcal{C}}(d) < 2^d$ for some $d \in \mathbf{N}$, and then we define its **Vapnik-Chervonenkis index** $V(\mathcal{C})$ to be the least such d; if $\mathcal{C}$ is not a VC-class we put $V(\mathcal{C}) = \infty$. By theorem (1.6), $V(\mathcal{C}) < \infty$ means that $f_{\mathcal{C}}$ is of polynomial growth, while $V(\mathcal{C}) = \infty$ means that $f_{\mathcal{C}}(n) = 2^n$ for all n.

(2.2) Usually a collection $\mathcal{C} \subseteq \mathcal{P}(X)$ will be indexed by elements of a parameter space or index set, and in fact the set X and this index set play dual roles. To bring out this duality, let us assume that infinite sets X and Y are given, and a binary relation $\Phi \subseteq X \times Y$. For $x \in X$ and $y \in Y$ we put

$$\Phi_x := \{y \in Y : (x,y) \in \Phi\}, \quad \Phi^y := \{x \in X : (x,y) \in \Phi\}.$$

Let $\Phi^Y := \{\Phi^y : y \in Y\} \subseteq \mathcal{P}(X)$ and $\Phi_X := \{\Phi_x : x \in X\} \subseteq \mathcal{P}(Y)$. (So Y is the parameter space for Φ^Y and X is the parameter space for Φ_X.)

Let $E \subseteq F \subseteq X$. Then we have for $y \in Y$ the obvious equivalence

$$E = \Phi^y \cap F \Leftrightarrow y \in \Phi_x \text{ for all } x \in E \text{ and } y \notin \Phi_x \text{ for all } x \in F - E,$$

hence

$$(*) \qquad E \in \Phi^Y \cap F \Leftrightarrow \left(\bigcap_{x \in E} \Phi_x \right) \cap \left(\bigcap_{x \in F-E} (Y - \Phi_x) \right) \ne \emptyset.$$

As E ranges over the subsets of a finite set $F \subseteq X$, the nonempty intersections on the right in $(*)$ are exactly the atoms of the boolean algebra of subsets of Y generated by the Φ_x with $x \in F$. So we conclude (for finite $F \subseteq X$):

$$(**) \qquad \left\{ \begin{array}{l} |\Phi^Y \cap F| = \text{number of atoms of the boolean algebra of} \\ \qquad\qquad \text{subsets of } Y \text{ generated by the sets } \Phi_x \text{ with } x \in F. \end{array} \right.$$

(2.3) This leads us to a situation dual to the earlier set-up: Fix a set Y. For $S \subseteq Y$ we write $S^1 := S$, and $S^{-1} := Y - S$ (the complement of S in Y). Given sets $S_1, \ldots, S_n \subseteq Y$ we note that the 2^n sets (many of which may be empty)

$$S_1^{\epsilon(1)} \cap \cdots \cap S_n^{\epsilon(n)}, \quad \epsilon: \{1, \ldots, n\} \to \{-1, 1\},$$

are disjoint and cover Y, so the boolean algebra $B(S_1, \ldots, S_n)$ of subsets of Y generated by $S_1, \ldots, S_n$ has as its atoms the *nonempty* sets among these intersections. Hence the maximal number of atoms is 2^n and this is achieved precisely when all 2^n intersections indicated are nonempty. In this case we say that the sequence $S_1, \ldots, S_n$ is **independent** (in Y), or (abusing language) that $S_1, \ldots, S_n$ are independent (in Y). Otherwise we say that the sequence $S_1, \ldots, S_n$ is **dependent.** We also say that a *collection* $\mathcal{G}$ of subsets of Y is **independent** (in Y) if for each $n \in \mathbf{N}$ there is an independent sequence $S_1, \ldots, S_n$ in $\mathcal{G}$, and otherwise we call $\mathcal{G}$ **dependent**. The following is now dual to theorem (1.6).

(2.4) PROPOSITION. *Suppose $\mathcal{G} \subseteq \mathcal{P}(Y)$ is dependent, so there is d such that each sequence $S_1, \ldots, S_d$ in $\mathcal{G}$ is dependent. Then for such d and all $S_1, \ldots, S_n \in \mathcal{G}$, the boolean algebra $B(S_1, \ldots, S_n)$ has at most $p_d(n)$ atoms.*

PROOF. Let $n \geq d$ and $S_1, \ldots, S_n \in \mathcal{G}$. Let $F := \{1, \ldots, n\}$, and let $\mathcal{D} \subseteq \mathcal{P}(F)$ consist of the sets $D \subseteq F$ for which the intersection

$$\left(\bigcap_{i \in D} S_i \right) \cap \left(\bigcap_{i \notin D} S_i^{-1} \right)$$

is nonempty. If $|\mathcal{D}| > p_d(n)$ then by lemma (1.3) there would be a set $E \subseteq \{1, \ldots, n\}$ of size d such that $\mathcal{P}(E) = \mathcal{D} \cap E$, hence the boolean algebra generated by the sets S_i with $i \in E$ would have 2^d atoms, contradicting the assumption on d. □

(2.5) The simplest examples of dependent collections are obtained as follows. Let $\mathcal{B}$ be a collection of subsets of Y and d a positive integer such that each nonempty intersection $B_1 \cap \cdots \cap B_d$ of d sets from $\mathcal{B}$ equals an intersection $\bigcap_{i \in D} B_i$ for some proper subset D of the index set $\{1, \ldots, d\}$. For instance, this is the case for $\mathcal{B} =$ the set of connected subsets of the real line $\mathbb{R}$, with $d = 3$. More generally, given an o-minimal structure $(R, <, \mathcal{S})$, we take $Y = R$ and $\mathcal{B} =$ the collection of definably connected subsets of R; then $\mathcal{B}$ satisfies the property above with $d = 3$, as is easily checked.

(2.6) LEMMA. *Let $\mathcal{B}$ and d be as in the general statement of (2.5). Let $\mathcal{G} \subseteq \mathcal{P}(Y)$ and suppose $e \in \mathbf{N}$ is such that each set in $\mathcal{G}$ is a boolean combination of at most e sets in $\mathcal{B}$. Then $\mathcal{G}$ is dependent. More precisely, $B(S_1, \ldots, S_n)$ has at most $p_d(en)$ atoms, for all $S_1, \ldots, S_n \in \mathcal{G}$.*

PROOF. Let each S_i be a boolean combination of sets $B_{i1}, \ldots, B_{ie}$ in $\mathcal{B}$, and let $F := \{(i, j) : 1 \leq i \leq n,\ 1 \leq j \leq e\}$, an index set of en elements. Then the boolean algebra generated by the S_i's is contained in the boolean algebra generated by the B_λ's with $\lambda \in F$, and the latter boolean algebra has as atoms the nonempty sets among the intersections

$$\left(\bigcap_{\lambda \in D} B_\lambda\right) \cap \left(\bigcap \left\{B_\mu^{-1} : \mu \in F,\ B_\mu \text{ does not contain } \bigcap_{\lambda \in D} B_\lambda\right\}\right)$$

where D varies over the subsets of F of size $< d$. Hence there are at most $p_d(en)$ atoms in the boolean algebra generated by the B_λ's with $\lambda \in F$, and hence at most $p_d(en)$ atoms in the boolean algebra generated by $S_1, \ldots, S_n$. □

(2.7) EXAMPLE. Let $(R, <, \mathcal{S})$ be an o-minimal structure and $\Phi \subseteq R^{m+1}$ a definable set and put $\mathcal{G} := \{\Phi_x : x \in R^m\}$, a collection of subsets of R. By Chapter 3, (3.6) we know that for some fixed $e \in \mathbf{N}$ each set in $\mathcal{G}$ has at most e definably connected components. Hence by taking for $\mathcal{B}$ the collection of definably connected subsets of R the lemma shows that $\mathcal{G}$ is dependent, and that the boolean algebra generated by sets $S_1, \ldots, S_n \in \mathcal{G}$ has at most $p_3(en)$ atoms. Later we shall see that for each definable set $\Phi \subseteq R^{m+n}$ the collection $\Phi_X := \{\Phi_x : x \in R^m\}$ of subsets of R^n is dependent.

(2.8) Given an infinite set Y and a collection $\mathcal{G} \subseteq \mathcal{P}(Y)$ we define a "growth function" $f^{\mathcal{G}} : \mathbf{N} \to \mathbf{N}$ by

$$f^{\mathcal{G}}(n) := \text{maximum over all } S_1, \ldots, S_n \in \mathcal{G} \text{ of the number of atoms of } B(S_1, \ldots, S_n).$$

In the case where $\mathcal{G}$ is dependent, we let $D(\mathcal{G})$ be the smallest $d \in \mathbf{N}$ for which $f^{\mathcal{G}}(d) < 2^d$, and we call $D(\mathcal{G})$ the **dependency index** of $\mathcal{G}$. If $\mathcal{G}$ is independent we set $D(\mathcal{G}) = \infty$.

(2.9) *Suppose X and Y are infinite sets and $\Phi \subseteq X \times Y$ is a binary relation.*

Then we defined in (2.2) above the collections $\mathcal{C} = \Phi^Y \subseteq \mathcal{P}(X)$ and $\mathcal{G} = \Phi_X \subseteq \mathcal{P}(Y)$, and (*) and (**) show that $f_{\mathcal{C}} = f^{\mathcal{G}}$, hence $V(\Phi^Y) = D(\Phi_X)$, and in particular,

$$\Phi^Y \text{ is a VC-class} \Leftrightarrow \Phi_X \text{ is dependent},$$

and by reversing the roles of $\mathcal{C}$ and $\mathcal{G}$,

$$\Phi^Y \text{ is dependent} \Leftrightarrow \Phi_X \text{ is a VC-class}.$$

Less obvious perhaps is

(2.10) PROPOSITION. *Φ^Y is independent if and only if Φ_X is independent.*

PROOF. By symmetry we need only prove one direction, so assume Φ^Y is independent, and let $n \in \mathbf{N}$. We want to find $x(1), \ldots, x(n) \in X$ such that $\Phi_{x(1)}, \ldots, \Phi_{x(n)} \subseteq$

Y are independent. We know there are 2^n independent sets $\Phi^{y(\epsilon)}$ ($\epsilon \in \{-1,1\}^n$) in X, so in particular there is for each $m \in \{1,\ldots,n\}$ a point $x(m)$ in X with

$$x(m) \in \left(\bigcap_{\epsilon(m)=1} \Phi^{y(\epsilon)} \right) \cap \left(\bigcap_{\epsilon(m)=-1} \left(\Phi^{y(\epsilon)}\right)^{-1} \right),$$

that is, $(x(m), y(\epsilon)) \in \Phi \Leftrightarrow \epsilon(m) = 1$, for all $\epsilon \in \{1,-1\}^n$, hence

$$y(\epsilon) \in \Phi_{x(m)} \Leftrightarrow \epsilon(m) = 1,$$

and thus

$$y(\epsilon) \in \bigcap_{1 \le m \le n} \left(\Phi_{x(m)}\right)^{\epsilon(m)}.$$

Therefore $\Phi_{x(1)}, \ldots, \Phi_{x(n)}$ are independent. □

(2.11) Because of this proposition and the equivalences in (2.9) the properties "Φ^Y is a VC-class", "Φ^Y is dependent", "Φ_X is a VC-class", "Φ_X is dependent" are all equivalent. The proof of (2.10) shows moreover

$$D(\Phi^Y) > 2^n \Rightarrow D(\Phi_X) > n,$$

hence

$$D(\Phi_X) \le n \Rightarrow D(\Phi^Y) \le 2^n,$$

or in terms of $C = \Phi^Y$ alone,

$$D(C) \le 2^{V(C)}.$$

We call the binary relation Φ **dependent** or **independent**, according to whether Φ_X is dependent or independent, and we set $D(\Phi) := D(\Phi_X) = V(\Phi^Y)$.

A useful source of VC-classes is the following result from Dudley [24].

(2.12) PROPOSITION. *Let $\mathcal{L}$ be an m-dimensional real vector space of real-valued functions on an infinite set X, and for each $f \in \mathcal{L}$, put* $\operatorname{pos}(f) := \{x \in X : f(x) > 0\}$. *Then* $\operatorname{pos}(\mathcal{L}) := \{\operatorname{pos}(f) : f \in \mathcal{L}\}$ *is a VC-class of subsets of X with* $V(\operatorname{pos}(\mathcal{L})) = m + 1$.

PROOF. Let $A \subseteq X$ be a set with $|A| = m + 1$. We shall see that then not each subset of A is cut out from A by a set in $\operatorname{pos}(\mathcal{L})$. The restriction map $f \mapsto f|A : \mathcal{L} \to \mathbb{R}^A$ cannot be surjective since $\dim(\mathbb{R}^A) = m+1 > \dim(\mathcal{L})$. Therefore $\mathbb{R}^A$ contains a nonzero vector w that is orthogonal to all $f|A$ ($f \in \mathcal{L}$) with respect to the standard inner product $(u, v) := \sum_{a \in A} u(a) \cdot v(a)$ on $\mathbb{R}^A$. Replacing w by $-w$ if necessary we may as well assume that $A^+ := \{a \in A : w(a) > 0\}$ is nonempty. If there were $f \in \mathcal{L}$ with $A^+ = A \cap \operatorname{pos}(f)$, we would have $0 = (w, f|A) = \sum w(a) \cdot f(a) > 0$, a contradiction. Hence A^+ is not cut out from A by any set in $\operatorname{pos}(\mathcal{L})$. This shows

$V(\text{pos}(\mathcal{L})) \leq m+1$, and we leave the proof that $V(\text{pos}(\mathcal{L}))$ is not less than $m+1$ as an exercise. □

(2.13) In particular the proposition applies to the vector space of real polynomial functions on $X = \mathbb{R}^N$ of degree $\leq d$, where N is a positive integer. In the exercises below this leads to a quite elementary proof that every semialgebraic binary relation $\Phi \subseteq \mathbb{R}^M \times \mathbb{R}^N$ $(M, N > 0)$ is dependent. (This is also a special case of the main theorem of this chapter.)

(2.14) EXERCISES.

1. Finish the proof of (2.12) by showing that $V(\text{pos}(\mathcal{L})) \geq m+1$.

Let X, Y be infinite sets. Given $\Phi \subseteq X \times Y$, put $f^{\Phi} := f^{\mathcal{G}} : \mathbf{N} \to \mathbf{N}$, where $\mathcal{G} = \Phi_X$, so $f^{\Phi}(n) =$ the maximum over all $x(1), \ldots, x(n)$ in X of the number of atoms of the boolean algebra $B(\Phi_{x(1)}, \ldots, \Phi_{x(n)})$.

Let $\Phi, \Psi \subseteq X \times Y$, and define the relations $\neg\Phi$, $\Phi \vee \Psi$, $\Phi \& \Psi \subseteq X \times Y$ as follows: $(x,y) \in \neg\Phi$ iff $(x,y) \notin \Phi$, $(x,y) \in \Phi \vee \Psi$ iff $(x,y) \in \Phi$ or $(x,y) \in \Psi$, $(x,y) \in \Phi \& \Psi$ iff $(x,y) \in \Phi$ and $(x,y) \in \Psi$. Note that $f^{\Phi} = f^{\neg\Phi}$.

2. Show that $f^{\Phi \vee \Psi} \leq f^{\Phi} \cdot f^{\Psi}$, and derive that $f^{\Phi \& \Psi} \leq f^{\Phi} \cdot f^{\Psi}$.

3. Show that if Φ and Ψ are dependent, then $\Phi \vee \Psi$ and $\Phi \& \Psi$ are dependent.

4. Derive from the previous two exercises and proposition (2.12) that every semialgebraic relation $\Phi \subseteq \mathbb{R}^M \times \mathbb{R}^N$ $(M, N > 0)$ is dependent.

§3. Reduction to the case $q = 1$

In this section we prove the following rather difficult result due to Shelah [53]:

(3.1) THEOREM. *Let $\mathcal{R} = (R, \ldots)$ be an infinite model-theoretic structure and suppose all definable relations $\Phi \subseteq R^p \times R$, for all $p > 0$, are dependent. Then all definable relations $\Phi \subseteq R^p \times R^q$, for all $p, q > 0$, are dependent.*

(3.2) Following Laskowski [38] we establish this in a purely combinatorial way, via a somewhat more precise result than (3.1), namely theorem (3.12) below. As an essential tool we first prove a well-known combinatorial fact due to Ramsey.

(3.3) Given a set X and a positive integer r, we put

$$X^{(r)} := \text{the collection of all } r\text{-element subsets of } X.$$

(3.4) RAMSEY'S THEOREM. *Given positive integers M, r and k there is a positive integer $N = N(M, r, k)$ so large that if X is a set with $|X| \geq N$ and $X^{(r)} = P_1 \cup P_2 \cup \cdots \cup P_k$, then there is a set $Y \subseteq X$ with $|Y| = M$ such that $Y^{(r)} \subseteq P_j$ for some $j \in \{1, \ldots, k\}$.*

PROOF. Clearly it suffices to prove this for $k = 2$, and we do this case by induction on r. If $r = 1$ we can take $N = 2M$. Assume true for a certain r, so we have positive integers $N(M, r, 2)$ with the desired property. Next define positive integers $N(1), \ldots, N(2M)$ by descending recursion:

$$N(2M) = 1 \text{ and } N(i) := N\big(N(i+1), r, 2\big) + 1 \text{ for } 1 \leq i < 2M.$$

We show that then $N = N(M, r+1, 2) := N(1)$ has the required property. Let $X = \{1, \ldots, N\}$, with $N = N(1)$ as above, and let $X^{(r+1)} = P_1 \cup P_2$. We construct a descending sequence $A_1 \supseteq A_2 \supseteq \cdots \supseteq A_{2M}$ of subsets of X, with $|A_i| \geq N(i)$, and a sequence $a_1, \ldots, a_{2M}$, with $a_i \in A_i$ for $i = 1, \ldots, 2M$, and $a_i \notin A_{i+1}$ for $i < 2M$. To start with, put $A_1 = X = \{1, \ldots, N\}$, and let a_1 be any element of A_1. Suppose A_i and $a_i \in A_i$ have been constructed, $|A_i| \geq N(i)$, $1 \leq i < 2M$. Then $(A_i - \{a_i\})^{(r)} = Q_1 \cup Q_2$, with

$$Q_1 := \big\{\{b_1, \ldots, b_r\} \in (A_i - \{a_i\})^{(r)} : \{a_i, b_1, \ldots, b_r\} \in P_1\big\},$$
$$Q_2 := \big\{\{b_1, \ldots, b_r\} \in (A_i - \{a_i\})^{(r)} : \{a_i, b_1, \ldots, b_r\} \in P_2\big\}.$$

Since $|A_i - \{a_i\}| \geq N\big(N(i+1), r, 2\big)$, the inductive hypothesis implies there is $A_{i+1} \subseteq A_i - \{a_i\}$ with $|A_{i+1}| \geq N(i+1)$ and $A_{i+1}^{(r)} \subseteq Q_1$ or $A_{i+1}^{(r)} \subseteq Q_2$. Let a_{i+1} be any element of A_{i+1}. This finishes the construction of the A_i's and a_i's. Let $A = \{a_1, \ldots, a_{2M}\}$, and note that for each $i \in \{1, \ldots, 2M\}$,

$$\text{either (1) } \{a_i, a_{i(1)}, \ldots, a_{i(r)}\} \in P_1 \text{ whenever } i < i(1) < \cdots < i(r) \leq 2M,$$
$$\text{or (2) } \{a_i, a_{i(1)}, \ldots, a_{i(r)}\} \in P_2 \text{ whenever } i < i(1) < \cdots < i(r) \leq 2M.$$

Let $Y_1 := \{a_i : \text{ (1) holds}\}$ and $Y_2 := \{a_i : \text{ (2) holds}\}$. Then $A = Y_1 \cup Y_2$, and either $|Y_1| \geq M$ or $|Y_2| \geq M$, and clearly $Y_1^{(r+1)} \subseteq P_1$ and $Y_2^{(r+1)} \subseteq P_2$. So either $Y = Y_1$ or $Y = Y_2$ has the required property. □

(3.5) Ramsey's theorem is intimately related to the model-theoretic notion of indiscernibility, which we need here only in a very simple form, as follows.

DEFINITIONS. Let X be an infinite set. Given a relation $A \subseteq X^r$ we say that a finite sequence $x_1, \ldots, x_M$ in X is **A-indiscernible** if

$$\big(x_{i(1)}, \ldots, x_{i(r)}\big) \in A \Leftrightarrow \big(x_{j(1)}, \ldots, x_{j(r)}\big) \in A.$$

whenever $1 \leq i(1) < \cdots < i(r) \leq M$ and $1 \leq j(1) < \cdots < j(r) \leq M$. Let $\mathcal{A}$ be a collection of relations on X, that is, each element of $\mathcal{A}$ is a set $A \subseteq X^r$, with $r \in \mathbf{N}$ depending on A. Then a sequence $x_1, \ldots, x_M$ in X is called **$\mathcal{A}$-indiscernible** if it is A-indiscernible for each $A \in \mathcal{A}$. A **subsequence** of a sequence $x_1, \ldots, x_N$ is by definition a sequence $x_{i(1)}, \ldots, x_{i(M)}$ with $1 \leq i(1) < \cdots < i(M) \leq N$.

(3.6) COROLLARY. *Let X be an infinite set and $\mathcal{A}$ a finite collection of relations on X. Then, given any positive integer M there is a positive integer N so large that each sequence in X of length N contains an $\mathcal{A}$-indiscernible subsequence of length M.*

PROOF. By an induction on $|\mathcal{A}|$ it suffices to prove this when $\mathcal{A}$ consists of just one relation $A \subseteq X^r$, $r > 0$. Let $x_1, \ldots, x_N$ be a sequence in X of length N, with $N = N(M, r, 2)$, and write $\{1, \ldots, N\}^{(r)} = P_1 \cup P_2$ with

$$P_1 := \{\{i(1), \ldots, i(r)\} : 1 \leq i(1) < \cdots < i(r) \leq N, \ (x_{i(1)}, \ldots, x_{i(r)}) \in A\},$$
$$P_2 := \{\{i(1), \ldots, i(r)\} : 1 \leq i(1) < \cdots < i(r) \leq N, \ (x_{i(1)}, \ldots, x_{i(r)}) \notin A\}.$$

By Ramsey's theorem there is $\{i(1), \ldots, i(M)\} \subseteq \{1, \ldots, N\}$, with $i(1) < \cdots < i(M)$, such that $\{i(1), \ldots, i(M)\}^{(r)} \subseteq P_1$ or $\{i(1), \ldots, i(M)\}^{(r)} \subseteq P_2$. Then the subsequence $x_{i(1)}, \ldots, x_{i(M)}$ of $x_1, \ldots, x_N$ is clearly A-indiscernible. $\square$

(3.7) For the rest of this section we fix infinite sets X and Y and a binary relation $\Phi \subseteq X \times Y$.

(3.8) LEMMA. *Suppose Φ is independent, and $\mathcal{A}$ is a finite collection of relations on X. Then there are for each $M \in \mathbf{N}$ an $\mathcal{A}$-indiscernible sequence $a_1, \ldots, a_M$ in X and an element $b \in Y$ such that for all $m \in \{1, \ldots, M\}$, $(a_m, b) \in \Phi \Leftrightarrow m$ is even.*

PROOF. Fix a natural number M. By (3.6) we can find a natural number N so large that each sequence in X of length N contains an $\mathcal{A}$-indiscernible subsequence of length M. As Φ is independent there are elements $x(i) \in X$ for $1 \leq i \leq N$ such that the sets $\Phi_{x(1)}, \ldots, \Phi_{x(N)}$ are independent; so for each set $w \subseteq \{1, \ldots, N\}$ the intersection

$$\left(\bigcap_{i \in w} \Phi_{x(i)}\right) \cap \left(\bigcap_{i \notin w} (\Phi_{x(i)})^{-1}\right) \tag{1}$$

is nonempty.

Fix an $\mathcal{A}$-indiscernible subsequence $x(i_1), \ldots, x(i_M)$ of length M of $(x(i))_{1 \leq i \leq N}$, and take for b an element of the intersection (1) for $w := \{i_{2m} : 1 \leq m \leq M/2\}$. Then the lemma above holds for the sequence $a_1, \ldots, a_M$ where $a_m := x(i_m)$. $\square$

(3.9) Next we introduce certain relations that are in some sense "definable" from Φ. Let the variables $x_1, \ldots, x_M$ range over X and the variable y over Y.

Given a positive integer M and a set $u \subseteq \{1, \ldots, M\}$ we put

$$\Phi_u(x_1, \ldots, x_M; y) := \left(\bigwedge_{i \in u} \Phi(x_i, y)\right) \ \& \ \left(\bigwedge_{i \notin u} \neg\Phi(x_i, y)\right),$$

a formula defining a subset Φ_u of $X^M \times Y$; the image of Φ_u under the projection map $X^M \times Y \to X^M$ is defined by the formula $\exists y \Phi_u(x_1, \ldots, x_M; y)$, and here we simply regard this last formula as a convenient notation for the subset of X^M it defines. So

$$\mathcal{A}_{\Phi,M} := \{\exists y \Phi_u(x_1, \ldots, x_M; y) : u \subseteq \{1, \ldots, M\}\}$$

is simply a finite collection of M-ary relations on X.

The following is a converse to the previous lemma.

(3.10) LEMMA. *Let $a(1), \ldots, a(N)$ be a sequence in X that is $\mathcal{A}_{\Phi,M}$-indiscernible, $N \geq 2M$, let $a(i_1), \ldots, a(i_{2M})$ be a subsequence, and let $b \in Y$ be such that for all $m \in \{1, \ldots, 2M\}$, $(a(i_m), b) \in \Phi \Leftrightarrow m$ is even. Then $D(\Phi) > M$.*

PROOF. It suffices to show that $\Phi_{a(1)}, \ldots, \Phi_{a(M)}$ are independent. Note first that $\Phi_E(a(i_1), \ldots, a(i_{2M}); b)$ holds for $E :=$ set of even integers in $\{1, \ldots, 2M\}$. Given any set $u \subseteq \{1, \ldots, M\}$, take a sequence $1 \leq k(1) < \cdots < k(M) \leq 2M$ such that $k(i)$ is even for $i \in u$ and $k(i)$ is odd for $i \notin u$. Then clearly

$$\Phi_u(a(i_{k(1)}), \ldots, a(i_{k(M)}); b)$$

holds. Hence $\exists y \Phi_u(a(i_{k(1)}), \ldots, a(i_{k(M)}); y)$ holds, and thus

$$\exists y \Phi_u(a(1), \ldots, a(M); y)$$

holds by indiscernability. Since $u \subseteq \{1, \ldots, M\}$ was arbitrary this shows that $\Phi_{a(1)}, \ldots, \Phi_{a(M)}$ are independent. □

(3.11) Next we assume also that the set Y is given as a cartesian product $Y = Y_1 \times Y_2$, with both Y_1 and Y_2 infinite. Let the variable y_1 range over Y_1 and y_2 over Y_2. Consider now the formula $\Phi(x; y_1, y_2)$ as defining a relation $\Phi^* \subseteq (X \times Y_1) \times Y_2$. (This is just Φ if we identify $(X \times Y_1) \times Y_2$ with $X \times (Y_1 \times Y_2)$ in the usual way.) The relation Φ^* parametrizes a collection of subsets of Y_2 with index set $X \times Y_1$, and so it makes sense to say that Φ^* is dependent.

Given a positive integer M and a set $u \subseteq \{1, \ldots, M\}$ we introduce a formula

$$\Gamma_{\Phi,u}(x_1, \ldots, x_M; y_1) := \exists y_2 \Phi_u(x_1, \ldots, x_M; y_1, y_2).$$

This formula $\Gamma_{\Phi,u}$ defines a subset of $X^M \times Y_1$, and we will simply regard $\Gamma_{\Phi,u}$ as a notation for this subset. So $\Gamma_{\Phi,u}$ parametrizes a collection of subsets of Y_1 with index set X^M. So it makes sense to say that $\Gamma_{\Phi,u}$ is dependent. With these notations we have

(3.12) THEOREM. *Suppose there are positive integers M and N such that $D(\Phi^*) \leq M$ and $D(\Gamma_{\Phi,u}) \leq N$ for all $u \subseteq \{1, \ldots, M\}$. Then Φ is dependent.*

REMARK. The constructive nature of the proof below and of the arguments above would in principle allow us to extract a (huge) bound $D(\Phi) \leq B(M, N)$, for some explicit function B of M and N. The construction of this function B would involve the bounds in Ramsey's theorem.

PROOF. We introduce a formula $\Psi_{u,v}$ for $u \subseteq \{1,\ldots,M\}$ and $v \subseteq \{1,\ldots,N\}$:

$$\Psi_{u,v}(x^1,\ldots,x^N) := \exists y_1 \left(\bigwedge_{j \in v} \Gamma_{\Phi,u}(x^j; y_1) \ \& \ \bigwedge_{j \notin v} \neg\Gamma_{\Phi,u}(x^j; y_1) \right),$$

where each $x^j = (x^{j_1},\ldots,x^{j_M})$ is a variable ranging over X^M. So $\Psi_{u,v}$ defines a subset of $(X^M)^N = X^{MN}$. Actually $\Psi_{u,v}$ is just $\exists y_1(\Gamma_{\Phi,u})_v(x^1,\ldots,x^N; y_1)$ in the notation of (3.9) above. Put $\mathcal{A}_u := \{\Psi_{u,v} : v \subseteq \{1,\ldots,N\}\}$ and

$$\mathcal{A} := \{\Psi_{u,v} : u \subseteq \{1,\ldots,M\},\ v \subseteq \{1,\ldots,N\}\},$$

so $\mathcal{A}$ is the union of the $\mathcal{A}_u$'s.

Suppose Φ is independent. (We shall derive a contradiction.) Put

$$K := (2N)^{2^M} \cdot 2M.$$

By lemma (3.8) there are an $\mathcal{A}$-indiscernible sequence $a_1,\ldots,a_K$ in X and an element $b = (b_1,b_2) \in Y_1 \times Y_2$ such that $\Phi(a_k; b)$ holds if and only if k is even. Since $D(\Phi^*) \leq M$ there is by lemma (3.10) no interval J of length $2M$ contained in $\{1,\ldots,K\}$ so that the sequence $(a_k,b_1)_{k \in J}$ is $\mathcal{A}_{\Phi^*,M}$-indiscernible.

Each relation in $\mathcal{A}_{\Phi^*,M}$ is defined by a formula of the form

$$\exists y_2 \Phi^*_u((x_1,y_{11}),\ldots,(x_M,y_{1M}); y_2)$$

with $u \subseteq \{1,\ldots,M\}$ and each variable (x_i,y_{1i}) ranging over $X \times Y_1$. For $(a_k,b_1)_{k \in J}$ to be indiscernible with respect to this relation is equivalent to $(a_k)_{k \in J}$ being $\Gamma_{\Phi,u}(x_1,\ldots,x_M; b_1)$-indiscernible. (Here $\Gamma_{\Phi,u}(x_1,\ldots,x_M; b_1)$ stands for the subset of X^M defined by it.) So there is no interval J of length $2M$ contained in $\{1,\ldots,k\}$ such that $(a_k)_{k \in J}$ is $\Gamma_{\Phi,u}(x_1,\ldots,x_m; b_1)$-indiscernible for all $u \subseteq \{1,\ldots,M\}$.

CLAIM. Let P and Q be positive integers with $Q \geq 2NP$, let $I = \{k : i_0 \leq k < i_0 + Q\}$ be an interval of length Q contained in $\{1,\ldots,K\}$, and let $u \subseteq \{1,\ldots,M\}$. Then I has a subinterval $J := \{k : j_0 \leq k < j_0 + P\}$ of length P, such that the sequence $(a_k)_{k \in J}$ is $\Gamma_{\Phi,u}(x_1,\ldots,x_M; b_1)$-indiscernible.

Note that if the claim holds, then starting with $Q := K$ and $P := Q/2N$, and successively applying the claim to each of the 2^M subsets $u \subseteq \{1,\ldots,M\}$ we obtain in the end an interval J of length $2M$ contained in $\{1,\ldots,K\}$ so that the sequence $(a_k)_{k \in J}$ is $\Gamma_{\Phi,u}(x_1,\ldots,x_M; b_1)$-indiscernible for all $u \subseteq \{1,\ldots,M\}$, and we have a contradiction as desired. So all that remains is to prove the claim.

PROOF OF CLAIM. Assume there is no such interval J. Now, given any j with $0 \leq j < 2N$ the interval $J(j) := \{k : i_0 + jP \leq k < i_0 + (j+1)P\}$ is of length P and contained in I, so in particular $(a_k)_{k \in J(j)}$ is not $\Gamma_{\Phi,u}(x_1, \ldots, x_M; b_1)$-indiscernible. So for some strictly increasing sequence of length M in $J(j)$ the corresponding sequence of a's does not satisfy $\Gamma_{\Phi,u}(x_1, \ldots, x_M; b_1)$. When j is even we choose a strictly increasing sequence $k(j,1) < \cdots < k(j,M)$ in $J(j)$ of the first kind, while for odd j we choose $k(j,1) < \cdots < k(j,M)$ in $J(j)$ of the second kind. Hence $\Gamma_{\Phi,u}\big(a_{k(j,1)}, \ldots, a_{k(j,M)}; b_1\big)$ holds if and only if j is even, for $0 \leq j < 2N$. Put $a_j^* := \big(a_{k(j,1)}, \ldots, a_{k(j,M)}\big) \in X^M$ for $0 \leq j < 2N$. Since $(a_k)_{k \in I}$ is an $\mathcal{A}$-indiscernible sequence, the order in which the a_k's appear within an "increasing" sequence of a_j^*'s implies that $\big(a_j^*\big)_{0 \leq j < 2N}$ is $\mathcal{A}_u$-indiscernible. But $\Gamma_{\Phi,u}\big(a_j^*; b_1\big)$ holds if and only if j is even, which by lemma (3.10) contradicts $D(\Gamma_{\Phi,u}) \leq N$. □

(3.13) We call attention to the fact that for dependence of Φ we need only dependence of the auxiliary relations Φ^* and $\Gamma_{\Phi,u}$, which are defined purely in terms of Φ; also, the definitions of Φ^* and $\Gamma_{\Phi,u}$ in terms of Φ do not involve quantifiers over X.

(3.14) PROOF OF THEOREM (3.1). We are given an infinite model-theoretic structure $\mathcal{R} = (R, \ldots)$ such that each definable relation $\Phi \subseteq R^p \times R$, for each $p > 0$, is dependent. We have to show that all definable relations $\Phi \subseteq R^p \times R^q$ are dependent, for all $p, q > 0$. The case $q = 1$ holds by assumption. Assume inductively that the desired property holds for a certain $q > 0$, and let $\Phi \subseteq R^p \times R^{q+1}$ be definable, $p > 0$. Now set $X := R^p$, $Y_1 := R^q$, and $Y_2 := R$, $Y := Y_1 \times Y_2 = R^{q+1}$. Then $\Phi \subseteq X \times Y$, so we are in the setting of (3.11) above. By the hypothesis of the theorem the (definable) relation $\Phi^* \subseteq R^{p+q} \times R$ is dependent, say $D(\Phi^*) \leq M$, for a certain positive integer M. Since clearly all relations $\Gamma_{\Phi,u} \subseteq R^{Mp} \times R^q$ with $u \subseteq \{1, \ldots, M\}$ are definable, the inductive assumption implies there is some positive integer N such that $D(\Gamma_{\Phi,u}) \leq N$ for all $u \subseteq \{1, \ldots, M\}$. Then theorem (3.12) tells us that Φ is dependent, as desired. □

(3.15) COROLLARY. *Let $(R, <, \mathcal{S})$ be an o-minimal structure and $\Phi \subseteq R^p \times R^q$ $(p, q > 0)$ a definable relation. Then Φ is dependent. In particular, there is $d = d(\Phi) \in \mathbb{N}$ such that for all sufficiently large n each n-element set $F \subseteq R^q$ has at most n^d subsets of the form $\Phi_x \cap F$ with $x \in R^p$.*

PROOF. From (2.7) it follows that $(R, <, \mathcal{S})$ satisfies the hypothesis of theorem (3.1). Hence Φ is dependent. Thus $\Phi_X = \{\Phi_x : x \in R^p\}$ is a VC-class, by (2.11). □

Notes and comments

The combinatorial dichotomy of Section 1 is due to Shelah [53], and independently to Vapnik and Chervonenkis [62]. In [53] this had to do with the model-theoretic "independence property", and in this connection Shelah also proved (3.1) by a curious set-theoretic argument.

In [62] the motivation came from probability theory: if $\mathcal{C}$ is a VC-class of events in a probability space X, then the estimates in Bernoulli's theorem for the events in $\mathcal{C}$ are uniform in a certain technical sense. See also Dudley [24] for more on this.

That o-minimal structures do not have the independence property was shown by Pillay and Steinhorn [49]. The combinatorial interpretation in terms of the VC-property was provided by Laskowski [38] who also gave the purely combinatorial proof of (3.1) in Section 3. Wilkie (unpublished) gave another proof of (3.1) for o-minimal structures using their special properties. Somewhat earlier Stengle and Yukich [58] had shown (as in (2.13) and (2.14) above) that semialgebraic collections have the VC-property.

For recent applications to the study of neural networks, see Macintyre and Sontag [41] and Sontag [56].

CHAPTER 6

POINT-SET TOPOLOGY IN O-MINIMAL STRUCTURES

Introduction

In this chapter we fix an o-minimal expansion $(R, <, \mathcal{S})$ of an ordered abelian group $(R, <, 0, -, +)$. (As was shown in Chapter 1, this group operation makes R an abelian divisible torsion-free group, so we may and do consider R as a vector space over $\mathbb{Q}$.)

For convenience we fix a positive element $1 \in R$. We also set

$$|x| := \begin{cases} x \text{ if } x \geq 0, \\ -x \text{ if } x < 0. \end{cases}$$

The presence of such a group operation leads in Section 1 to quick proofs of curve selection and of the fact that a definable continuous map on a closed bounded definable set has closed bounded image.

Definable curves here play a role similar to that of sequences in $\mathbb{R}^n$, but have better properties. Section 2 proves that "fiberwise open" implies "piecewise open" for definable sets $S \subseteq R^{m+n}$, and variants of this important fact. In Section 3 we show that "definably connected" equals "definably path connected", and we obtain some results on definable partitions of unity. In Section 4 we consider definably proper and definably identifying maps.

§1. Curve selection

(1.1) An important consequence of the presence of the group structure is that we can definably pick an element $e(X) \in X$ from each nonempty definable set X. The idea is to let $e(X)$ be the midpoint of X if X is a bounded interval. Here are the details:

(i) Let $X \subseteq R$ be definable and nonempty. If X has a least element, then we let $e(X)$ be this least element. If X does not have a least element, let (a, b) be its "left-most" interval: $a = \inf X$, $b = \sup\{x \in R : (a, x) \subseteq R\}$. Then

$a < b$ and $(a,b) \subseteq X$; now put

$$e(X) := \begin{cases} 0 & \text{if } a = -\infty, b = +\infty, \\ b-1 & \text{if } a = -\infty,\ b \in R, \\ a+1 & \text{if } a \in R,\ b = +\infty, \\ (a+b)/2 & \text{if } a, b \in R. \end{cases}$$

(ii) Let $X \subseteq R^m$ be definable and nonempty, $m > 1$, and let $\pi : R^m \to R^{m-1}$ be the projection on the first $m-1$ coordinates. Then $\pi X \subseteq R^{m-1}$ so we may assume inductively that an element $a = e(\pi X)$ of πX has been defined. Then $X_a \subseteq R$ and we put $e(X) := \big(a, e(X_a)\big)$.

(1.2) PROPOSITION (DEFINABLE CHOICE).

(i) *If $S \subseteq R^{m+n}$ is definable and $\pi : R^{m+n} \to R^m$ the projection on the first m coordinates, then there is a definable map $f : \pi S \to R^n$ such that $\Gamma(f) \subseteq S$.*

(ii) *Each definable equivalence relation on a definable set X has a definable set of representatives.*

PROOF. For (i), define $f(x) := e(S_x)$ for $x \in \pi S$. For (ii) note that

$$\{e(A) : A \text{ is an equivalence class}\}$$

is a definable set of representatives. □

(1.3) COMMENT. Model-theorists will recognize in (i) the property of having definable Skolem functions. As to (ii), consider our structure $\mathcal{S}$ on R as a category with the definable sets as objects and the definable maps between them as morphisms. If $E \subseteq X \times X$ is a definable equivalence relation on a definable set $X \subseteq R^m$, then (ii) provides a definable map $f : X \to X$ such that $E = \text{kernel}(f)$, that is, $xEy \Leftrightarrow f(x) = f(y)$, for all $x, y \in X$. Hence for each definable map $g : X \to R^n$ such that $xEy \Rightarrow g(x) = g(y)$ for $x, y \in X$, there is a unique definable map $g' : f(X) \to R^n$ such that $g = g' \circ f$. Therefore the definable set $f(X)$, together with the map $f : X \to f(X)$, serves as a quotient X/E in the category $\mathcal{S}$.

(1.4) We equip from now on R^m, $m > 0$, with the "supnorm" $|\ |$:

$$|x| = \max\{|x_1|, \ldots, |x_m|\} \text{ for } x = (x_1, \ldots, x_m).$$

Note that then $|\ | : R^m \to R$ is a definable continuous function. We also define $|\ | : R^0 = \{0\} \to R$ by $|0| = 0$.

(1.5) COROLLARY (CURVE SELECTION). *If $a \in \text{cl}(X) - X$, where X is definable, then there is a definable continuous injective map $\gamma : (0, \epsilon) \to X$, for some $\epsilon > 0$, such that $\lim_{t \to 0} \gamma(t) = a$.*

PROOF. Since $a \in \text{cl}(X) - X$, the definable set $\{|a - x| : x \in X\} \subseteq R$ contains arbitrarily small positive elements, hence contains an interval $(0, \epsilon)$, $\epsilon > 0$. For each

t in this interval there is $x \in X$ with $|a - x| = t$. By definable choice there is then a definable map $\gamma : (0, \epsilon) \to X$ such that $|a - \gamma(t)| = t$ for all $t \in (0, \epsilon)$. By decreasing ϵ if necessary we may assume by the monotonicity theorem that γ is continuous. Obviously γ is injective and $\lim_{t \to 0} \gamma(t) = a$. □

(1.6) Definable curve selection fails in the o-minimal structure $(\mathbb{R}, <)$ (see exercise 8 at the end of this section), so our standing assumption in this chapter that we are dealing with an o-minimal expansion of an ordered group is appropriate here. A point in the closure of a set X in a metric space is the limit of a sequence in X. This fact is not available in our context, but curve selection offers a more than adequate alternative (curves instead of sequences). We shall use curve selection in this way to prove that the image of a closed bounded definable set under a continuous definable map is closed and bounded. Here a set $A \subseteq R^m$ is called **bounded** if $|a| < r$ for all a in A and a fixed $r \in R$. First some lemmas.

(1.7) LEMMA. *Let C be a bounded cell in R^m, $m > 1$, and $\pi : R^m \to R^{m-1}$ the projection on the first $m-1$ coordinates. Then $\pi\,\mathrm{cl}(C) = \mathrm{cl}(\pi C)$.*

PROOF. We shall do the case $C = (f, g)_{\pi C}$ and leave the other case to the reader. By the continuity of π we have $\pi\,\mathrm{cl}(C) \subseteq \mathrm{cl}(\pi C)$. Let $a \in \mathrm{cl}(\pi C)$. We have to find s in R such that $(a, s) \in \mathrm{cl}(C)$. Of course we may assume $a \notin \pi C$. Then there is a continuous definable map $\gamma : (0, \epsilon) \to \pi C$ such that $\lim_{t \to 0} \gamma(t) = a$. We are going to lift the curve γ to a curve in C. Since C is bounded there is $r > 0$ such that $-r < f(x) < g(x) < r$ for all $x \in \pi C$. Define the continuous function $\lambda : (0, \epsilon) \to R$ by $\lambda(t) := \big(f(\gamma(t)) + g(\gamma(t))\big)/2$. Note that $-r < \lambda(t) < r$ for all t, so by the monotonicity theorem there is $s \in R$ such that $\lim_{t \to 0} \lambda(t) = s$. Then $t \mapsto \big(\gamma(t), \lambda(t)\big) : (0, \epsilon) \to C$ is a continuous definable map whose limit as t goes to 0 equals (a, s), so $(a, s) \in \mathrm{cl}(C)$. □

REMARK. We cannot omit the hypothesis that C is bounded: in the ordered field of real numbers, consider $C := \big\{(x, 1/x) \in \mathbb{R}^2 : x > 0\big\}$.

(1.8) Next we recall that if $f : X \to Y$ is a continuous map from a topological space X into a Hausdorff space Y, then its graph $\Gamma(f)$ is a closed subset of $X \times Y$.

(1.9) LEMMA. *Let $f : X \to R^n$ be a definable continuous map on a closed bounded set $X \subseteq R^m$. Then $f(X)$ is bounded in R^n.*

PROOF. Suppose $\forall t \in R\ \exists x \in X\ |f(x)| > t$. By definable choice there is then a definable map $g : R \to X$ such that $\big|f\big(g(t)\big)\big| > t$ for all t in R. Since X is closed and bounded it follows from the monotonicity theorem, applied to the m coordinates of g, that $\lim_{t \to \infty} g(t) = x$ exists and belongs to X. So $f(x) = f\big(\lim_{t \to \infty} g(t)\big) = \lim_{t \to \infty} f\big(g(t)\big)$, but the last limit cannot exist in R, since $\big|f\big(g(t)\big)\big| > t$ for all t. Contradiction. □

(1.10) PROPOSITION. *If $f : X \to R^n$ is a continuous definable map on a closed bounded set $X \subseteq R^m$, then $f(X)$ is closed and bounded in R^n.*

PROOF. By (1.9) we already know that $f(X)$ is bounded in R^n, so it suffices to show that $f(X)$ is closed. Consider $Y := \{(f(x), x) : x \in X\} \subseteq R^{n+m}$ (the "reversed" graph of f) and write

$$Y = C_1 \cup \cdots \cup C_k,$$

where $C_1, \ldots, C_k$ are cells in R^{n+m}. Since Y is closed by (1.8) we also have

$$Y = \mathrm{cl}(C_1) \cup \cdots \cup \mathrm{cl}(C_k).$$

By (1.9) the set Y is bounded, so the cells $C_1, \ldots, C_k$ are bounded. Then it follows from (1.7) that

$$\begin{aligned} \pi Y &= \pi\,\mathrm{cl}(C_1) \cup \cdots \cup \pi\,\mathrm{cl}(C_k) \\ &= \mathrm{cl}(\pi C_1) \cup \cdots \cup \mathrm{cl}(\pi C_k), \end{aligned}$$

where $\pi : R^{n+m} \to R^n$ is the projection on the first n coordinates. But $\pi Y = f(X)$, so $f(X)$ is closed in R^n. □

Here are three easy consequences, the first dealing with the case $n = 1$.

(1.11) COROLLARY. *If $f : X \to R$ is a continuous definable function on a nonempty closed bounded set $X \subseteq R^m$, then f assumes a maximum and a minimum value.*

(1.12) COROLLARY. *If $f : X \to R^n$ is an injective continuous definable map on a closed bounded set $X \subseteq R^m$, then f is a homeomorphism from X onto $f(X)$.*

This last corollary perhaps requires explanation: f maps *definable* closed subsets of X onto closed subsets of $f(X)$, by (1.10), hence it maps *definable* open subsets of X onto open subsets of $f(X)$; now use the fact that an arbitrary open subset of X is a union of definable open subsets of X.

(1.13) COROLLARY. *Let $f : X \to R^n$ be a definable continuous map on a closed bounded set $X \subseteq R^m$ and let $Y = f(X)$. Then we have:*

(i) *A definable set $S \subseteq Y$ is closed if and only if $f^{-1}(S)$ is closed;*
(ii) *A definable map $g : Y \to R^p$ is continuous if and only if $g \circ f : X \to R^p$ is continuous.*

PROOF. Assertion (i) is clear from (1.10) and (ii) is an easy consequence of (i), taking into account that for continuity of g it suffices that $g^{-1}(Z)$ is closed for all definable closed $Z \subseteq R^p$. □

Closed bounded definable sets have a property that is like the completeness property of projective varieties in algebraic geometry:

(1.14) PROPOSITION. *Let $X \subseteq R^m$ be a closed bounded definable set and $Y \subseteq R^n$ any definable set. Then the projection map $X \times Y \to Y$ maps definable closed subsets of $X \times Y$ onto definable closed subsets of Y.*

PROOF. Let $q: X \times Y \to Y$ be the projection map, let $A \subseteq X \times Y$ be definable and closed in $X \times Y$, and suppose $y \in \mathrm{cl}_Y(q(A))$. Then there is for each $t > 0$ in R a point $a \in A$ such that $|q(a) - y| < t$. By definable choice and the monotonicity theorem this gives a definable continuous map $\alpha : (0, \epsilon) \to A$, for some $\epsilon > 0$, such that $|q(\alpha(t)) - y| < t$ for $t \in (0, \epsilon)$. Write $\alpha(t) = (\beta(t), \gamma(t))$, where β and $\gamma = q \circ \alpha$ are definable continuous maps from $(0, \epsilon)$ into X and Y respectively, so $\lim_{t \to 0} \gamma(t) = y$. Since X is bounded, it follows from Chapter 3, (1.6) that $x := \lim_{t \to 0} \beta(t)$ exists in R^m; hence $x \in X$, since X is closed. Then $(x, y) \in X \times Y$ and $\lim_{t \to 0} \alpha(t) = (x, y)$, so $(x, y) \in A$, because A is closed in $X \times Y$. Therefore $y = q(x, y) \in q(A)$ as desired. □

(1.15) EXERCISES.

In the first three exercises we consider a definable equivalence relation $E \subseteq X \times X$ on a definable set $X \subseteq R^m$.

1. Show that the definable set of representatives indicated in the proof of (1.2) is definable in the model-theoretic structure $(R, <, 1, +, E)$. (Recall that this set of representatives is given by $T := \{e(A) : A \text{ is an equivalence class}\}$.)

2. Show that E has only finitely many equivalence classes of dimension $\dim(X)$, and that each of them is definable in the model-theoretic structure $(R, <, 1, +, E)$.

3. Suppose all equivalence classes of E have the same Euler characteristic e. Show that then the Euler characteristic of X is a multiple of e. (In particular, this shows that for $e > 1$ there is no definable equivalence relation on R^m all of whose equivalence classes have exactly e elements.)

4. (Uniform continuity) Let $X \subseteq R^m$ be a closed and bounded definable set and $f: X \to R^n$ a continuous definable map. Show that there is for each $\epsilon > 0$ a $\delta > 0$ such that whenever $|x - y| < \delta$, $x, y \in X$, we have $|f(x) - f(y)| < \epsilon$.

5. (Fixed point theorem) Let X be a nonempty closed bounded definable subset of R^m and $f: X \to X$ a definable map such that $|f(x) - f(y)| < |x - y|$ for all distinct points $x, y \in X$. Show that f has a unique fixed point.

6. (Uniform curve selection) Let $X \subseteq R^m$ be definable. Show there are definable maps $\epsilon : \partial X \to (0, \infty)$ and $\Gamma : (0, \epsilon) \to X$ such that for each $a \in \partial X$ the function $t \to \Gamma(a, t): (0, \epsilon(a)) \to X$ is continuous, injective and satisfies $\lim_{t \to 0} \Gamma(a, t) = a$.

7. Given a map $f: A \to R^n$, $A \subseteq R^m$, we call f **locally bounded** if each point $a \in A$ has a neighborhood U in A such that $f(U)$ is bounded. Let $A \subseteq R^m$ be definable and $f: A \to R^n$ definable. Prove the following equivalence:

$$f \text{ is continuous} \iff f \text{ is locally bounded and } \Gamma(f) \text{ is closed in } A \times R^n .$$

Note that the local boundedness condition cannot be omitted in the reverse implication $\Leftarrow$: the function $f:\mathbb{R}\to\mathbb{R}$ given by $f(x) = 1/x$ for $x \neq 0$, $f(0) = 0$, is definable in the ordered field of real numbers, has closed graph $\Gamma(f)$ in $\mathbb{R}^2$, but is not continuous at 0.

8. Consider the o-minimal model-theoretic structure $(\mathbb{R}, <)$ and the set

$$X := \{(x,y)\in\mathbb{R}^2 : 0 < x < 1,\ 0 < y < 2\},$$

which is definable in $(\mathbb{R}, <)$ using the constants $0, 1, 2$. Note that $(1,2) \in \mathrm{cl}(X)$ and show that there is no subset Y of X such that Y is definable in $(\mathbb{R}, <)$ using constants, $\dim(Y) = 1$ and $(1,2)\in\mathrm{cl}(Y)$.

§2. Fiberwise properties

(2.1) Strictly speaking the results of this section are not needed later in this book. However, for those familiar with elementary model theory this section provides a very cheap way to obtain the triviality theorem of Chapter 9 from the triangulation theorem of Chapter 8, see Chapter 8, (2.13) and (2.14), for more details. The proof of the theorem below nicely illustrates the use of the definable choice principle.

(2.2) THEOREM. *Let $S \subseteq R^{m+n}$ be a definable set such that for each $x \in R^m$ the fiber S_x is open in R^n. Then there is a partition of R^m into cells $C_1,\dots,C_k$ such that $S \cap (C_i \times R^n)$ is open in $C_i \times R^n$ for $i = 1,\dots,k$. (Same for "closed" instead of "open".)*

PROOF. By induction on m. The case $m = 0$ is trivial. Let $m > 0$ and assume the result holds for lower values of m. First we show:

$$(*)\qquad \begin{cases} \text{Each open cell } C \text{ in } R^m \text{ has an open subcell } D \\ \text{such that } S\cap(D\times R^n) \text{ is open in } D\times R^n. \end{cases}$$

To see this, note that $C = C(1)\cup C(2)$ with definable $C(i)$ as follows:

$$\begin{aligned} C(1) &:= \{x\in C:\ (\{x\}\times S_x)\cap \mathrm{cl}((C\times R^n)-S)\neq\emptyset\},\\ C(2) &:= \{x\in C:\ (\{x\}\times S_x)\cap \mathrm{cl}((C\times R^n)-S)=\emptyset\}. \end{aligned}$$

If $C(2)$ has nonempty interior, we can take for D any open cell contained in $C(2)$. Now assume $C(2)$ has empty interior. Then $C(1)$ has nonempty interior, and replacing C by an open subcell contained in $C(1)$ we reduce to the case that $C = C(1)$. So for each $x\in C$ there is a point $s(x)\in S_x$ with $(x, s(x))\in\mathrm{cl}((C\times R^n)-S)$. Since S_x is open in R^n there is also for each $x \in C$ an $\epsilon(x) > 0$ such that if $y\in R^n$ and $|y - s(x)| < \epsilon(x)$, then $y\in S_x$. By definable choice we may take $x\mapsto s(x): C\to R^n$ and $x\mapsto\epsilon(x): C\to R$ to be definable functions: replacing C by a suitable open subcell we may further reduce to the case that these functions are continuous.

Then $\{(x,y) \in C \times R^n : x \in C, |y - s(x)| < \epsilon(x)\}$ is a subset of S, and open in $C \times R^n$; it also contains the points $(x, s(x))$ $(x \in C)$, which belong to the closure of $(C \times R^n) - S$, a blatant contradiction. This proves $(*)$.

Now let A be the union of all open boxes D in R^m such that $S \cap (D \times R^n)$ is open in $D \times R^n$. Then A is definable, $S \cap (A \times R^n)$ is open in $A \times R^n$, and $(*)$ implies that $\dim(R^m - A) < m$. Write $R^m - A$ as a disjoint union of cells $B_1, \ldots, B_r$ where each B_i has dimension $< m$.

Let B be one of the cells B_i, and let $\dim(B) = d < m$. Then $p(B)$ is an open cell in R^d, and we can use the homeomorphism $(x,y) \mapsto (p_B(x), y) : B \times R^n \to p(B) \times R^n$ and apply the inductive hypothesis to the image of $S \cap (B \times R^n)$ under this homeomorphism. The "closed" version follows by passing to complements. □

(2.3) Corollary. *Let $S' \subseteq S$ be definable sets in R^{m+n}, and let $A \subseteq R^m$ be definable such that S'_a is open in S_a for all $a \in A$. Then there is a partition of A into definable subsets $A_1, \ldots, A_M$ such that $S' \cap (A_i \times R^n)$ is open in $S \cap (A_i \times R^n)$, for $i = 1, \ldots, M$. (Same with "closed" instead of "open".)*

Proof. In this case it is more convenient to do the closed version. Replacing S by $S \cap (A \times R^n)$ we may as well assume that the projection map $R^{m+n} \to R^m$ maps S into A; thus S'_x is closed in S_x for each $x \in R^m$. Let S^* be the subset of R^{m+n} with $S^*_x = \text{cl}(S'_x)$ for each $x \in R^m$. Then S^* is definable and $S^*_x \cap S_x = S'_x$ for $x \in R^m$. Now apply the closed version of the theorem above to S^*. □

(2.4) Corollary. *Let $S \subseteq R^{m+n}$ be definable, $f : S \to R^k$ a locally bounded definable map, and $A \subseteq R^m$ a definable set such that for all $a \in A$ the map $f_a : S_a \to R^k$ is continuous, where $f_a(y) := f(a,y)$. Then there is a partition of A into definable subsets $A_1, \ldots, A_M$ such that each restriction*

$$f|S \cap (A_i \times R^n) : S \cap (A_i \times R^n) \to R^k$$

is continuous.

Proof. Apply the closed version of (2.3) to the definable subset $\Gamma(f)$ of $S \times R^k$, noting that $\Gamma(f)_a = \Gamma(f_a)$ is closed in $S_a \times R^k = (S \times R^k)_a$ for each $a \in A$. This gives a partition of A into definable subsets $A_1, \ldots, A_M$ such that for each $i \in \{1, \ldots, M\}$ the set $\Gamma(f) \cap (A_i \times R^{n+k}) = \Gamma(f|S \cap (A_i \times R^n))$ is closed in $(S \times R^k) \cap (A_i \times R^{n+k}) = (S \cap (A_i \times R^n)) \times R^k$. Hence, by (1.15), exercise 7, the restrictions $f|S \cap (A_i \times R^n)$ are continuous. □

(2.5) Exercises.

1. Assume $(R, <, \mathcal{S})$ expands an ordered field. Show that then (2.4) holds even without the assumption that f is locally bounded.

2. Assume $(R, <, \mathcal{S})$ expands an ordered field. Let $S \subseteq R^{m+n}$ be definable, $f : S \to R^k$ a definable map, and $A \subseteq R^m$ a definable set such that $f|S \cap (A \times R^n)$ is injective and $f_a : S_a \to R^k$ is a homeomorphism from S_a onto $f_a(S_a)$ for all $a \in A$. Show that there is a partition of A into definable subsets $A_1, \ldots, A_M$ such that each restriction $f|S \cap (A_i \times R^n) : S \cap (A_i \times R^n) \to R^k$ is a homeomorphism from $S \cap (A_i \times R^n)$ onto $f\big(S \cap (A_i \times R^n)\big)$.

§3. Paths and partitions of unity

The two topics of this section are unrelated. The results here will be essential in several later chapters.

(3.1) Definable paths.

Let $X \subseteq R^m$. A **definable path** in X is a definable continuous map $\gamma : [a, b] \to R^m$ with $a, b \in R$, $a < b$, taking values in X. Such a path γ is said to connect the points $\gamma(a)$ and $\gamma(b)$. If $\gamma : [a, b] \to X$ and $\delta : [b, c] \to X$ are definable paths in X with $\gamma(b) = \delta(b)$ we have a definable path $\gamma \vee \delta : [a, c] \to X$ in X agreeing on $[a, b]$ with γ and on $[b, c]$ with δ. Using this concatenation construction it is clear that if the points $x, y \in X$ can be connected by a definable path in X, and also the points $y, z \in X$, then x, z as well. It is easy to see that if X is definable and every two points in X can be connected by a definable path in X, then X is definably connected. The converse fails for the o-minimal model-theoretic structure $(\mathbb{R}, <)$ (see the exercise at the end of this section). It does hold however under our standing assumption in this chapter that we are dealing with an o-minimal expansion of an ordered abelian group:

(3.2) Proposition. *Suppose the definable set X is definably connected. Then any two points in X can be connected by a definable path in X.*

Proof. Assume first that X is a cell, without loss of generality an open cell in R^m. This case is handled by induction on m. The case $m = 1$ is obvious. For $m > 1$, let C be the projection of X in R^{m-1}, so $X = (f, g)$ with f, g functions on C, and let (y, r) and (z, s) be two points in X, $y, z \in C$. Assume f and g are R-valued. (The case that $f = -\infty$ or $g = +\infty$ is left to the reader.)

We first connect (y, r) by a "vertical" path in X to $(y, (f(y) + g(y))/2)$; by the inductive hypothesis there is a definable path $\gamma : [a, b] \to C$ connecting y to z and this path lifts to the path $[a, b] \to X$ given by $t \to \big(\gamma(t), (f(\gamma(t)) + g(\gamma(t)))/2\big)$, connecting $\big(y, (f(y) + g(y))/2\big)$ to $\big(z, (f(z) + g(z))/2\big)$; the last point can in turn be connected to (z, s) by a vertical path in X. Concatenating the three paths we obtain a definable path in X connecting (y, r) to (z, s).

In the general case, we use Chapter 3, (2.19), exercise 5 to write X as a union of cells $C_1, \ldots, C_k$ where for each $i < k$, either C_i intersects the closure of C_{i+1}, or

C_{i+1} intersects the closure of C_i. Then by curve selection there is a definable path in $C_i \cup C_{i+1}$ that connects a point of C_i with a point in C_{i+1}. Now combine this with the fact that the desired result has already been proved for cells. □

One particular consequence of (3.2) will be used in Chapters 8 and 9 in proofs of triangulation and trivialization:

(3.3) COROLLARY. *Let X and Y be definable sets in R^n with X definably connected, and $X \cap \mathrm{bd}(Y) = \emptyset$. Then either $X \subseteq Y$ or $X \cap Y = \emptyset$.*

PROOF. If not, there would be points $x \in X - Y$ and $y \in X \cap Y$. Take a definable path $\gamma : [a, b] \to X$ connecting x to y. Put $c := \inf\{t \in [a, b] : \gamma(t) \in Y\}$. Then $\gamma(c) \in X \cap \mathrm{bd}(Y)$, contradiction. □

PARTITIONS OF UNITY.

(3.4) LEMMA. *Let $A \subseteq B \subseteq R^m$ be definable sets, with A closed in B. Then there is a definable continuous function $f : B \to [0,1]$ with $f^{-1}(0) = A$.*

PROOF. Assuming $A \neq \emptyset$, take $f(x) = \min(1, d(x, A))$, where

$$d(x, A) = \inf\{|x - a| : a \in A\}. \quad \square$$

(3.5) LEMMA. *Let A_0 and A_1 be disjoint definable closed subsets of the definable set $B \subseteq R^m$. Then there are disjoint definable open subsets U_0 and U_1 of B with $A_i \subseteq U_i$, $i = 0, 1$.*

PROOF. Take functions f_0 and f_1 as in (3.4), and put

$$U_0 := \{x \in B : f_0(x) < f_1(x)\}, \text{ and } U_1 := \{x \in B : f_1(x) < f_0(x)\}. \quad \square$$

(3.6) LEMMA (SHRINKING OF OPEN COVERINGS). *Let the definable set $B \subseteq R^m$ be the union of its definable open subsets $U_1, \ldots, U_n$. Then B is also the union of definable open subsets $V_1, \ldots, V_n$ with $\mathrm{cl}_B(V_i) \subseteq U_i$ for $i = 1, \ldots, n$.*

PROOF. Assume inductively that $V_i \subseteq U_i$ has been defined for $i = 1, \ldots, k$ $(k < n)$ such that V_i is definable, open in B, $\mathrm{cl}_B(V_i) \subseteq U_i$ and $V_1, \ldots, V_k, U_{k+1}, \ldots, U_n$ cover B. Then apply (3.5) to the following two disjoint closed subsets of B:

$$A_0 := B - U_{k+1},$$

$$A_1 := B - \left(\bigcup_{i=1}^{k} V_i \cup \bigcup_{j=k+2}^{n} U_j \right). \quad \square$$

(3.7) LEMMA (DEFINABLE PARTITION OF UNITY). *With $U_1, \ldots, U_n$ and B as in (3.6), there are definable continuous functions $f_1, \ldots, f_n : B \to [0,1]$ such that*

(i) $\text{supp}(f_i) \subseteq U_i$ *for* $i = 1, \ldots, n$, *where* $\text{supp}(f_i) := \text{cl}_B(\{x \in B : f_i(x) \neq 0\})$,
(ii) $\sum f_i(x) > 0$ *for all* $x \in B$.

Moreover, if $(R, <, \mathcal{S})$ expands a real closed ordered field, then (ii) *can be replaced by $\sum f_i = 1$. (We then call $f_1, \ldots, f_n$ a definable partition of unity for the covering $U_1, \ldots, U_n$.)*

The easy proof using (3.6) and (3.4) is left to the reader.

(3.8) LEMMA. *Suppose $(R, <, \mathcal{S})$ expands an ordered real closed field. Let A_0 and A_1 be disjoint definable closed subsets of a definable set $B \subseteq R^m$. Then there is a continuous definable function $f : B \to [0,1]$ with $f^{-1}(0) = A_0$ and $f^{-1}(1) = A_1$.*

PROOF. Choose U_0 and U_1 as in (3.5), and apply (3.7) to get a definable partition of unity f_0, f_1, f_2 for the covering $U_0, U_1, B - (A_0 \cup A_1)$ of B. Then use (3.4) to get definable continuous functions $g_0 : B \to [0, 1/2]$ and $g_1 : B \to [1/2, 1]$ with $g_0^{-1}(0) = A_0$, $g_1^{-1}(1) = A_1$. Now set $f := f_0 g_0 + f_1 g_1 + (1/2) f_2$. □

(3.9) EXERCISE. Show that in the model-theoretic structure $(\mathbb{R}, <, 0, 1, 2)$ the definable set $\{(x, y) : 0 < x < 1,\ 0 < y < 2\} \cup \{(1, 2)\}$ in $\mathbb{R}^2$ is definably connected but not definably path connected (the latter defined as in (3.1)).

§4. Curves, proper maps, and identifying maps

This section contains miscellaneous results used only in the last chapter of this book. The reason for including these results here is simply that they hold under weaker assumptions than are in force later on in this book.

(4.1) DEFINITIONS. Let $X \subseteq R^m$ be a definable set.

(1) A **definable curve in** X is a definable map $\gamma : I \to X$, for some interval $I = (a, b)$ in R. We do not require γ to be continuous, and we allow of course that $b = +\infty$. Our interest is in the behavior of γ near its right endpoint b, that is in the germ of γ at b; we note that γ will be continuous on some smaller interval (a', b) with the same right endpoint. (That we are only interested in the behavior near the right endpoint is just a convention. We could as well have chosen the left endpoint.)
(2) Let $\gamma : (a, b) \to X$ be a definable curve in X. Given a point $p \in R^m$ we write $\gamma \to p$ as shorthand notation for $\lim_{t \to b} \gamma(t) = p$. (We do not require that $p \in X$.)

Let $X \subseteq R^m$ be definable and γ a definable curve in X. We call γ **completable** if there is a (necessarily unique) point $p \in R^m$ such that $\gamma \to p$. If $\gamma \to p \in X$, then we call γ **completable in** X. Here are some simple observations.

(i) X bounded $\Rightarrow$ γ is completable. (By Chapter 3, (1.6).)
(ii) X is closed and bounded $\Rightarrow$ γ is completable in X.
(iii) If $f: X \to Y$ is a definable map into a definable set $Y \subseteq R^n$, then $f(\gamma) := f \circ \gamma$ is a definable curve in Y; if f is in addition surjective ($f(X) = Y$), then each definable curve β in Y can be lifted to a definable curve α in X, that is, $f(\alpha) = \beta$. (By definable selection.)
(iv) γ is either injective on a subinterval (a', b), or constant on such a subinterval. (By the monotonicity theorem.)

(4.2) LEMMA. *Let $f: X \to Y$ be a definable map from the definable set $X \subseteq R^m$ into the definable set $Y \subseteq R^n$, and let $p \in X$. Then f is continuous at the point p iff for each definable curve γ in X with $\gamma \to p$ we have $f(\gamma) \to f(p)$.*

PROOF. If f is continuous at p, the "curve continuity" follows immediately. Suppose f is not continuous at p; so there is $\epsilon > 0$ such that the definable set $\{|x - p| : x \in X, |f(x) - f(p)| \geq \epsilon\}$ contains arbitrarily small positive elements, so it contains an interval I with left endpoint 0. By definable selection this produces a definable curve $\gamma: I \to X$ such that $|\gamma(t) - p| = t$ and $|f(\gamma(t)) - f(p)| \geq \epsilon$ for $t \in I$. Then its "transform" $\gamma': -I \to X$ given by $\gamma'(-t) = \gamma(t)$ has the property that $\gamma' \to p$ but not $f(\gamma') \to f(p)$. □

REMARK. As the proof shows, it suffices to consider just injective definable curves γ in the statement of the lemma.

(4.3) LEMMA. *Let the definable curve $\beta: (a, b) \to R^n$ be completable, $\beta \to p$, and let $\epsilon > 0$ be such that $a < b - \epsilon < b$ and β is continuous on $[b - \epsilon, b)$. Suppose $\alpha: I \to \beta([b - \epsilon, b))$ is a definable curve in $\beta([b - \epsilon, b))$ not completable in $\beta([b - \epsilon, b))$. Then α is a reparametrization of β near b: there are a subinterval I' of I with the same right endpoint as I, an $a' \in (b - \epsilon, b)$ and a definable strictly increasing homeomorphism $h: I' \to (a', b)$ such that $\alpha(t) = \beta(h(t))$ for $t \in I'$.*

PROOF. The curve α is necessarily completable in the closed and bounded set $\beta([b - \epsilon, b)) \cup \{p\}$. Hence $\alpha \to p$, and α and β are both injective near b. Thus by decreasing the interval I, keeping the same right endpoint, and also decreasing ϵ, we may as well assume that α is continuous and injective, and β is injective on $[b - \epsilon, b)$, and $p \notin \beta([b - \epsilon, b))$. Write $\alpha = (\alpha_1, \ldots, \alpha_n)$ and $\beta = (\beta_1, \ldots, \beta_n)$. By further shrinking we may also assume one of the β_i is strictly monotone on $[b - \epsilon, b)$, say β_1 is strictly increasing on $[b - \epsilon, b)$. Since $\alpha_1(I) \subseteq \beta_1([b - \epsilon, b))$, the function α_1 is also strictly increasing near the right endpoint of I.

Now use exercise 2 from Chapter 3, (1.9) to conclude that there are a subinterval I' of I with the same right endpoint as I, and $a' \in (b - \epsilon, b)$ and a definable strictly increasing homeomorphism $h: I' \to (a', b)$ such that $\alpha_1(t) = \beta_1(h(t))$ for $t \in I'$. Then it follows easily from $\alpha(I) \subseteq \beta([b - \epsilon, b))$ and the injectivity of β_1 on $[b - \epsilon, b)$ that also $\alpha(t) = \beta(h(t))$ for $t \in I'$. □

(4.4) DEFINITION. Let $f:X \to Y$ be a definable continuous map from the definable set $X \subseteq R^m$ into the definable set $Y \subseteq R^n$.

(1) We call f **definably proper** if for each definable set $K \subseteq Y$ we have: K is closed and bounded in $R^n \Rightarrow f^{-1}(K) \subseteq X$ is closed and bounded in R^m.

(2) We call f **definably identifying** if f is surjective ($f(X) = Y$), and for each definable set $K \subseteq Y$ we have: $f^{-1}(K)$ is closed in $X \Rightarrow K$ is closed in Y.

If $X \subseteq R^m$ is a closed and bounded definable set and $f: X \to R^n$ is a definable continuous map, then f is obviously definably proper, and, as a map from X onto $f(X)$, also definably identifying, by (1.10). A more typical example of a definably proper map is a projection map $R^m \times Y \to R^m$, where $Y \subseteq R^n$ is a closed and bounded definable set. These notions, and the lemmas that follow, will become especially useful in the last chapter when we construct quotient spaces of definable sets modulo definable equivalence relations.

(4.5) LEMMA. *Let $f:X \to Y$ be as above: X, Y definable, f definable and continuous. Then*

(1) *f is definably proper iff each definable curve γ in X whose image $f(\gamma)$ is completable in Y is completable in X;*

(2) *f is definably identifying iff each definable curve β in Y that is completable in Y lifts to a definable curve α in X that is completable in X.*

PROOF. (1) Suppose f is definably proper, $\gamma: (a,b) \to X$ is a definable curve in X and $f(\gamma)$ is completable in Y, say $f(\gamma) \to y \in Y$. Take $a' \in (a,b)$ such that γ is continuous on $[a', b)$. Then the definable set

$$K := \{f(\gamma(t)) : t \in [a', b)\} \cup \{y\} \subseteq Y$$

is closed and bounded in R^n, so $f^{-1}(K) \subseteq X$ is closed and bounded in R^m, so γ must be completable in $f^{-1}(K)$, hence in X.

Conversely, suppose f is not definably proper; then there is a definable $K \subseteq Y$ that is closed and bounded in R^n, such that $f^{-1}(K)$ is not closed in R^m or not bounded. If $f^{-1}(K)$ is not bounded, we can by definable selection choose a definable curve γ in $f^{-1}(K)$ that is not completable in R^m. Then $f(\gamma)$ is a curve in K, hence completable in K, and so in Y. If $f^{-1}(K)$ is not closed in R^m, take $p \in \text{cl}(f^{-1}(K)) - f^{-1}(K)$, and use curve selection to get a definable curve γ in $f^{-1}(K)$ such that $\gamma \to p$. Since $f^{-1}(K)$ is closed in X, we have $p \notin X$, so γ is not completable in X. But, as before, $f(\gamma)$ is completable in Y.

(2) Suppose f is definably identifying, and let $\beta:(a,b) \to Y$ be a definable curve in Y that is completable in Y. We have to lift β to a curve α in X that is completable in X. We may assume β is injective and continuous on a subset $[b-\epsilon, b)$ of (a,b), $\epsilon > 0$, since otherwise β would be constant on such a set, and we can then use the fact that f is surjective.

After further decreasing ϵ we may also assume that $\beta \to y \in Y - \beta([b-\epsilon, b))$, so $\beta([b-\epsilon, b))$ is not closed in Y. Hence $f^{-1}(\beta([b-\epsilon, b)))$ is not closed in X. By curve selection there is a definable curve α in $f^{-1}(\beta([b-\epsilon, b)))$ with $\alpha \to x$ and $x \in X - f^{-1}(\beta([b-\epsilon, b)))$. Then $f(\alpha)$ is a definable curve in $\beta([b-\epsilon, b))$ with $f(\alpha) \to f(x)$ and $f(x) \notin \beta([b-\epsilon, b))$. Hence by lemma (4.3) above, $f(\alpha)$ is a reparametrization of β near b. Reparametrizing α we achieve that $f(\alpha) = \beta$.

Conversely, suppose f is surjective but not definably identifying. Let $K \subseteq Y$ be definable such that $f^{-1}(K)$ is closed in X but K is not closed in Y. Then there is a definable curve β in K with $\beta \to y \in Y - K$. If β were to lift to a definable curve α in X that is completable in X, say $\alpha \to x \in X$, then image(α) $\subseteq f^{-1}(K)$ and $f^{-1}(K)$ would be closed in X, so $x \in f^{-1}(K)$, hence $y = f(x) \in K$, contradiction. □

(4.6) Corollary. *If $f: X \to Y$ as above is definably proper and surjective, then f is definably identifying.*

This is immediate from (4.5).

(4.7) Corollary. *Let $f: X \to Y$ as above be surjective. Then f is definably identifying iff for each definable $K \subseteq Y$ we have*

$$f(\mathrm{cl}_X(f^{-1}(K))) = \mathrm{cl}_Y(K).$$

Proof. Suppose f is definably identifying, and let $K \subseteq Y$ be definable. By continuity of f we have $f(\mathrm{cl}_X(f^{-1}(K))) \subseteq \mathrm{cl}_Y(K)$. For the reverse inclusion, let y be in the closure of K in Y, and take a definable curve β in K with $\beta \to y$. Lift β to a definable curve α in X with $\alpha \to x \in X$, so that $f(x) = y$. Then α lies in $f^{-1}(K)$, so $x \in \mathrm{cl}_X(f^{-1}(K))$, hence $y = f(x) \in f(\mathrm{cl}_X(f^{-1}(K)))$.

Conversely, assume each definable $K \subseteq Y$ has the property mentioned in the corollary. Let $\beta: (a, b) \to Y$ be a definable curve such that $\beta \to y \in Y$. We have to lift β to a definable curve in X completable in X. We may assume that β is not constant near b, hence we may assume β is injective and continuous on a subset $[b-\epsilon, b)$ of (a, b), $\epsilon > 0$, and $y \notin K := \beta([b-\epsilon, b))$. Then $y \in \mathrm{cl}_Y(K)$, so there is an x in $\mathrm{cl}_X(f^{-1}(K))$ with $f(x) = y$. Let α be a definable curve in $f^{-1}(K)$ with $\alpha \to x$. Then $f(\alpha)$ is a definable curve in K, hence the curve $f(\alpha)$ is a reparametrization of β near b, by (4.3). Then α can be similarly reparametrized so that $f(\alpha) = \beta$. □

(4.8) Exercises.

1. Let $f: X \to Y$ be a definably proper map. (This includes the assumption that X and Y are definable sets and f is definable and continuous.) Show that if $A \subseteq X$ is definable and closed in X, then $f(A)$ is closed in Y. Show that if $f': X' \to Y'$ is a definably proper map, then $f \times f': X \times X' \to Y \times Y'$ is definably proper. Show that if $g: Y \to Z$ is definably proper, then $g \circ f: X \to Z$ is definably proper.

2. Let $(R, <, \mathcal{S}')$ be an o-minimal structure with $\mathcal{S} \subseteq \mathcal{S}'$ and let $f : X \to Y$ be a definable continuous map between definable sets X and Y in R^m and R^n, where "definable" is taken in the sense of $\mathcal{S}$. Assume also that Y is locally closed in R^n. Show

$$f \text{ is definably proper with respect to } (R, <, \mathcal{S})$$
$$\Leftrightarrow$$
$$f \text{ is definably proper with respect to } (R, <, \mathcal{S}').$$

In the next exercise we assume that $(R, <, \mathcal{S}) = (\mathbb{R}, <, \mathcal{S})$ is an o-minimal expansion of the ordered additive group of reals. We recall that a continuous map $f : X \to Y$ between Hausdorff spaces X and Y is called **proper** if f maps closed subsets of X onto closed subsets of Y and each point $y \in Y$ has compact fiber $f^{-1}(y)$. This definition agrees with that of Bourbaki [6, Ch. 1, §10, Thm. 1] who also shows that if X and Y are Hausdorff spaces and the continuous map $f : X \to Y$ is proper, then $f^{-1}(C)$ is compact for each compact $C \subseteq Y$.

3. Let $f : X \to Y$ be a definable continuous map between definable sets $X \subseteq \mathbb{R}^m$ and $Y \subseteq \mathbb{R}^n$. Show that f is proper if and only if f is definably proper.

Notes and comments

The definable choice principle of Section 1 combines the model-theoretic properties of "definability of Skolem functions" and "elimination of imaginaries". The first use of definability of Skolem functions in the semialgebraic case seems to be in Delzell [14]. The stronger principle of definable choice was pointed out for real closed fields in [18]. That definability of Skolem function implies curve selection (for real and p-adic fields) was observed by Scowcroft and Van den Dries in [52]. Brumfiel makes the heuristically useful remark in [8] that curves play in semialgebraic geometry the same role as sequences in metric spaces. Peterzil and Steinhorn show in [47] that proposition (1.10) on images of closed bounded definable sets remains true for arbitrary o-minimal structures.

Bochnak, Coste, and Roy prove the fiberwise properties of Section 2 in the semialgebraic case in their book [4] by methods that seem specific to that situation. After the more general theorem (2.2) above had been found, Speissegger [57] established these properties for arbitrary o-minimal structures. The "partition of unity results" of Section 3 are taken over from the semialgebraic case treated by Delfs and Knebusch in [13, §1]. Similarly I adapted from Brumfiel [8] the material in Section 4 on curves, proper and identifying maps.

CHAPTER 7

SMOOTHNESS

Introduction

In this chapter we fix an o-minimal structure $(R, <, \mathcal{S})$ that expands an ordered field $(R, <, 0, 1, +, -, \cdot)$. Recall from Chapter 1, (4.6) that this ordered field is then necessarily real closed. In Section 1 we discuss differentiability, in Section 2 we show that definable functions in one variable are piecewise differentiable and that definable C^1-maps in several variables satisfy the implicit function theorem. This allows us to obtain in Section 3 the decomposition of definable sets into "smooth" cells. The smooth cell decomposition is then used in the last section to prove the good directions lemma, an important tool in triangulating definable sets in the next chapter.

§1. Differentiability in ordered fields

(1.1) We start by recording some elementary calculus facts valid over arbitrary ordered fields $(R, <, 0, 1, +, -, \cdot)$. (No o-minimality assumptions or definability assumptions on sets and functions are needed in this section.) To shorten notations we denote this ordered field just by R. We equip R with the interval topology, and each set R^m with the corresponding product topology, and we put $|x| := \sup\{|x_i| : i = 1, \ldots, m\}$ for $x = (x_1, \ldots, x_m) \in R^m$. We also equip R^m with the usual dot product: $x \cdot y = x_1 y_1 + \cdots + x_m y_m$. In this section we omit proofs since they are the same as for the field of real numbers.

(1.2) DEFINITION. Let $I \subseteq R$ be open. A function $f : I \to R^n$ is said to be **differentiable at a point $x \in I$ with derivative $a \in R^n$** if

$$\lim_{t \to 0} t^{-1}\big(f(x+t) - f(x)\big) = a.$$

Note that then f is continuous at x and that a is unique; we write $a = f'(x)$.

(1.3) Let the functions $f, g : I \to R^n$ be differentiable at x. Then the sum $f + g$ and the dot product $f \cdot g$ are differentiable at x and

$$\begin{aligned} (f+g)'(x) &= f'(x) + g'(x), \\ (f \cdot g)'(x) &= f'(x) \cdot g(x) + f(x) \cdot g'(x). \end{aligned}$$

If moreover $n = 1$ and g does not vanish on I then f/g is differentiable at x and

$$(f/g)'(x) = (f'(x) \cdot g(x) - f(x) \cdot g'(x))/g^2(x).$$

Constant maps are differentiable everywhere with derivative 0.

The identity function on R is differentiable everywhere with derivative 1.

Let I, J be open subsets of R, let $f: I \to R$ be continuous and differentiable at the point $x \in I$, $g: J \to R$ differentiable at $f(x) \in J$. Then $g \circ f$, defined on the open set $I \cap f^{-1}(J)$, is differentiable at x and

$$(g \circ f)'(x) = g'(f(x)) \cdot f'(x).$$

(1.4) DIRECTIONAL DERIVATIVES. Consider now a map $f: U \to R^n$ with $U \subseteq R^m$ open, a point $x \in U$, and a vector $v \in R^m$. We say that f **is differentiable at x in the v-direction with derivative $a \in R^n$** if the function $t \mapsto f(x + tv)$ (defined on an open neighborhood of $0 \in R^m$) is differentiable at 0 with derivative a, that is

$$\lim_{t \to 0} t^{-1} (f(x + tv) - f(x)) = a.$$

In that case we write $d_x f(v) = a$. Let $e(1), \ldots, e(m)$ be the standard basis of the R-linear space R^m, so $e(1) = (1, 0, \ldots, 0)$, etc. As in the case of the real field we write $(\partial f / \partial x_i)(x)$ for $d_x f(e(i))$ and call it the ith partial of f at x.

(1.5) THE DIFFERENTIAL OF A MAP. Let $f = (f_1, \ldots, f_n): U \to R^n$ be a map on an open subset U of R^m. Let $x \in U$ and $T: R^m \to R^n$ be a linear map.

We call f **differentiable at x with differential T** if for each $\epsilon > 0$ we have

$$|f(x + v) - f(x) - T(v)| < \epsilon |v|,$$

for all sufficiently small vectors $v \in R^m$. Then f is continuous at x and T is unique; we write $d_x f = T$. This notation is consistent with that in (1.4), since if T is as above, then f is differentiable at x in the v-direction for each v, and $T(v) =$ directional derivative of f at x in the v-direction. For $m = 1$ this notion of differentiability coincides with the one in (1.2), with $d_x f(1) = f'(x)$.

The map $f = (f_1, \ldots, f_n)$ is differentiable at x iff each coordinate function $f_i: U \to R$ is differentiable at x. In that case all partials $(\partial f_i / \partial x_j)(x)$ exist and the $n \times m$ matrix $(\partial f_i / \partial x_j)(x)$ is exactly the matrix of $d_x f$ relative to the standard bases. It is called the **Jacobian matrix of f at x.**

(1.6) BEHAVIOR UNDER ALGEBRAIC OPERATIONS. If the maps $f, g: U \to R^n$ are differentiable at $x \in U$, U open in R^m, then $f + g$ and cf, for each $c \in R$, are differentiable at x and

$$d_x(f + g) = d_x f + d_x g,$$
$$d_x cf = c \cdot d_x f.$$

If also $h: V \to R^p$ is differentiable at $f(x) \in V$, V open in R^n and f continuous, then $h \circ f$, defined on $U \cap f^{-1}V$, is differentiable at x and

$$d_x(h \circ f) = (d_{f(x)}h) \circ d_x f.$$

Each R-linear map $R^m \to R^n$ is differentiable at each point in R^m with itself as differential.

§2. Inverse function theorem

(2.1) CONVENTION FOR THE REST OF THIS CHAPTER.

$\mathcal{R} = (R, <, \mathcal{S})$ is an o-minimal expansion of an ordered field $(R, <, 0, 1, +, -, \cdot)$. (This ordered field is then in fact real closed by Chapter 1, (4.6).) We let $\|x\| := (x \cdot x)^{1/2}$ denote the usual 'euclidean norm' of the vector $x \in R^m$.

(2.2) LEMMA (ROLLE). *Suppose $a < b$ in R and the function $f : [a, b] \to R$ is definable, continuous, $f(a) = f(b)$, and f is differentiable at each point of (a, b). Then $f'(c) = 0$ for some $c \in (a, b)$.*

PROOF. Take c, $a < c < b$, such that $f(c)$ is maximal or minimal. Then clearly $f'(c) = 0$. □

This has the following easy consequences.

(2.3) MEAN VALUE THEOREM. *Suppose $a < b$ in R, $f : [a, b] \to R$ is definable and continuous, and differentiable at each point of (a, b). Then for some $c \in (a, b)$, we have $f(b) - f(a) = (b - a) \cdot f'(c)$.*

(2.4) THEOREM ON CONSTANTS. *Under the same assumptions as in the mean value theorem, suppose that moreover $f'(x) = 0$ for all $x \in (a, b)$. Then f is constant.*

In the following I denotes an interval. Our first goal is to prove

(2.5) PROPOSITION. *If $f : I \to R$ is definable, then f is differentiable at all but finitely many points of I.*

This requires several lemmas.

LEMMA 1. *Let $f : I \to R$ be definable. Then for each $x \in I$ the limits*

$$f'(x^+) := \lim_{t \downarrow 0} t^{-1}(f(x + t) - f(x)),$$
$$f'(x^-) := \lim_{t \uparrow 0} t^{-1}(f(x + t) - f(x))$$

exist in R_∞. If moreover f is continuous and $f'(x^+) > 0$ for all x, then f is strictly increasing and its inverse $f^{-1}: f(I) \to R$ satisfies $(f^{-1})'(y^+) = 1/f'(x^+)$, for $x \in I$ and $f(x) = y \in f(I)$. (Here $1/+\infty := 0$.)

PROOF. For fixed x the function $g: t \mapsto (f(x+t) - f(x))/t$, defined on an interval $(0, \epsilon)$, is definable, whence the limit $f'(x^+) = \lim_{t \to 0} g(t)$ exists by Chapter 3, (1.6). Similarly for $f'(x^-)$.

Suppose now that $f'(x^+) > 0$ for all x, and f is continuous. If f were not strictly increasing then f would be constant or strictly decreasing on some subinterval, contradicting $f'(x^+) > 0$ on that subinterval. The formula for the right derivative of f^{-1} is obtained in the usual way.

LEMMA 2. *Let $f: I \to R$ be definable and continuous, and suppose the maps $x \mapsto f'(x^+)$ and $x \mapsto f'(x^-)$ are R-valued and continuous on I. Then f is differentiable at each point of I, and $f': I \to R$ is continuous.*

PROOF. It suffices to show that $f'(a^+) = f'(a^-)$ for all $a \in I$. Suppose the contrary, say $f'(a^+) > f'(a^-)$ for a certain $a \in I$. Then there are $c \in R$ and a subinterval J of I around a such that $f'(x^+) > c > f'(x^-)$ on J. Hence the definable function $g: J \to R$ given by $g(x) := f(x) - cx$ has the property that $g'(x^+) > 0$, $g'(x^-) < 0$ for all x, so g would be both strictly increasing and strictly decreasing on J. Contradiction.

LEMMA 3. *Let $f: I \to R$ be definable. Then there are only finitely many $x \in I$ such that $f'(x) \in \{-\infty, +\infty\}$.*

PROOF. Suppose the definable set $\{x \in I : f'(x^+) = +\infty\}$ is infinite. Then this set contains a whole interval, and for the sake of deriving a contradiction we may as well assume that $f'(x^+) = +\infty$ for all $x \in I$ and that f is continuous. By lemma 1 this implies that f is strictly increasing, hence $f'(x^-) \geq 0$ for all $x \in I$.

After further shrinking the interval we may also assume that we are in one of the following cases:

(i) $f'(x^-) = +\infty$ for all x in I,
(ii) $f'(x^-) \in R$ for all x in I, and the map $x \mapsto f'(x^-)$ is continuous on I.

In case (i) the inverse of f satisfies $(f^{-1})'(y^-) = (f^{-1})'(y^+) = 0$ for all $y \in f(I)$, so that by the theorem on constants f^{-1} is constant, contradicting the injectivity of f^{-1}. In case (ii) we can apply the same argument as in the proof of lemma 2 to get a contradiction.

PROOF OF (2.5). By the monotonicity theorem and lemma 3 we can reduce to the case that f is continuous, and $f'(x^+)$ and $f'(x^-)$ are R-valued and continuous functions of $x \in I$. Now apply lemma 2. □

(2.6) Continuous differentiability.

In the following definition and lemmas we denote by $f = (f_1, \ldots, f_n): U \to R^n$ a *definable* map on a (definable) open set $U \subseteq R^m$. We also put

$$\mathrm{Lin}(R^m, R^n) \ := \ R\text{-linear space of } R\text{-linear maps } R^m \to R^n.$$

Definition. We call f a C^1**-map** if the partials $(\partial f_i / \partial x_j)$ are defined as R-valued functions on U and are continuous. (It follows that f is continuous.)

Lemma. *If f is C^1, then f is differentiable at each point of U, and the map*

$$x \mapsto d_x f : U \to \mathrm{Lin}(R^m, R^n) \cong R^{n \times m}$$

is continuous. Conversely, if f is differentiable at each point of U and the map $x \mapsto d_x f : U \to R^{n \times m}$ is continuous, then f is a C^1-map.

(Here we identify each R-linear map from R^m to R^n with its $n \times m$ matrix relative to the standard bases.)

The second part of the lemma is an immediate consequence of the fact stated at the end of (1.5). The proof of the first part follows the classical proof, which depends on the mean value theorem. (That is why we restrict ourselves to definable maps, for which this result is available.)

Next we define the norm of a linear map.

(2.7) Definition. For an R-linear map $T: R^m \to R^n$ we put

$$|T| \ := \ \max\{|Tx| : \ |x| \le 1, \ x \in R^m\}.$$

(Note that the maximum exists since T is definable and continuous and that $|T(x)| \le |T| \cdot |x|$ for all $x \in R^m$.)

(2.8) Lemma. *Let $f: U \to R^n$ be C^1, and let $[a, b] := \{(1-t)a + tb : \ 0 \le t \le 1\}$ be a line segment contained in U. Then*

$$|f(b) - f(a)| \le |b - a| \cdot \max_{y \in [a,b]} |d_y f|.$$

Proof. Define $g: [0,1] \to R^n$ by $g(t) := f\big((1-t)a + tb\big)$. Then g is differentiable at each t, $0 < t < 1$, with $g'(t) = d_y f(b-a)$, where $y := (1-t)a + tb$, so $|g'(t)| \le M$, with

$$M \ := \ |b - a| \cdot \max_{y \in [a,b]} d_y f.$$

Then clearly by the mean value theorem, $|f(b) - f(a)| = |g(1) - g(0)| \le M$. □

(2.9) LEMMA. *With the same assumptions as in the previous lemma, let $x \in U$. Then*

$$|f(b) - f(a) - d_x f(b-a)| \leq |b-a| \cdot \max_{y \in [a,b]} |d_y f - d_x f|.$$

PROOF. Apply the previous lemma to the map $y \mapsto f(y) - d_x f(y)$. □

(2.10) LEMMA. *With the same assumptions, let $m = n$, $a \in U$ and suppose $d_a f$ is invertible. Then there are $\epsilon > 0$ and $c > 0$ in R such that*

$$|f(x) - f(y)| \geq c|x-y| \text{ for all } x, y \in U \text{ with } |x-a|,\ |y-a| < \epsilon.$$

In particular f is injective on a neighborhood of a.

PROOF. Take $\epsilon > 0$ so small that the open ball $B(a, \epsilon)$ is contained in U. Let $x, y \in B(a, \epsilon)$. Applying the previous lemma, we have

$$|f(x) - f(y) - d_a f(x-y)| \leq |x-y| \cdot \max_{z \in [x,y]} |d_z f - d_a f|,$$

so

$$|f(x) - f(y)| \geq |d_a f(x-y)| - |x-y| \cdot \max_{z \in [x,y]} |d_z f - d_a f|.$$

Since $d_a f$ is invertible there is $c' > 0$, independent of x, y, such that

$$|d_a f(x-y)| \geq c'|x-y|.$$

(Use $|z| \geq c'|T(z)|$ for $T = (d_a f)^{-1}$, $z = d_a f(x-y)$.) Decreasing ϵ if necessary we may also assume that

$$|d_b f - d_a f| < c'/2 \text{ for all } b \in B(a, \epsilon).$$

Hence

$$\begin{aligned} |f(x) - f(y)| &\geq c'|x-y| - (c'/2)|x-y| \\ &\geq c|x-y| \end{aligned}$$

for $c = c'/2$. □

(2.11) INVERSE FUNCTION THEOREM. *Let $f : U \to R^m$ be a definable C^1-map on a definable open set $U \subseteq R^m$ and $a \in U$ a point where $d_a f : R^m \to R^m$ is invertible. Then there are definable open neighborhoods $U' \subseteq U$ of a and V' of $f(a)$ such that f maps U' homeomorphically onto V' and $f^{-1} : V' \to U'$ is also C^1.*

PROOF. In the statement of lemma (2.10) we may of course replace the norm $|\cdot|$ by the equivalent euclidean norm $||\cdot||$, which is more convenient here. Take $\epsilon, c > 0$ such that

$$\begin{aligned} ||x-a|| \leq \epsilon &\Rightarrow x \in U \text{ and } d_x f \text{ is invertible,} \\ \text{and } ||x-a||, ||y-a|| \leq \epsilon &\Rightarrow ||f(x) - f(y)|| \geq c||x-y||. \end{aligned}$$

In particular, $\|x-a\|=\epsilon \Rightarrow \|f(x)-f(a)\| \geq c\epsilon$. We claim that

$$\{y: \|y-f(a)\| < (1/2)c\epsilon\} \subseteq \{f(x): \|x-a\| < \epsilon\}.$$

To see this, let $\|y-f(a)\| < (1/2)c\epsilon$ and consider the function

$$P(x) := \|f(x)-y\|^2 = \sum (f_i(x)-y_i)^2 \text{ on the ball } \|x-a\| \leq \epsilon.$$

By Chapter 6, (1.11) the function P assumes a minimum value, but if $\|x-a\|=\epsilon$ then

$$\begin{aligned} P(x) &= \|(f(x)-f(a)) - (y-f(a))\|^2 > (c\epsilon - (1/2)c\epsilon)^2 \\ &= (1/4)c^2\epsilon^2 > \|y-f(a)\|^2 = P(a), \end{aligned}$$

so P must assume its minimum value at an interior point b, $\|b-a\| < \epsilon$. Then

$$0 = (\partial P/\partial x_j)(b) = \sum 2(f_i(b)-y_i)\cdot(\partial f_i/\partial x_j)(b) \text{ for all } j,$$

that is, $d_b f(f(b)-y) = 0$. Since $d_b f$ is invertible this gives $f(b)=y$, and we have proved the claim. Since ϵ can be taken arbitrarily small this argument shows that f maps each neighborhood of a onto a neighborhood of $f(a)$. Put $U' := \{x: \|x-a\| < \epsilon\}$. For the same reason f maps each neighborhood of each point $x \in U'$ onto a neighborhood of $f(x)$. Since f is also injective on U' it follows that f maps U' homeomorphically onto the open set $f(U')$. It remains to show that $f^{-1}: f(U') \to U'$ is a C^1-map.

Let $y \in f(U')$ approach $f(a)$ and put $f^{-1}(y) := b \in U'$, so b approaches a. From

$$(f(b)-f(a)-d_a f(b-a))/\|b-a\| \to 0 \text{ as } b \to a,$$

and $\|b-a\| \leq c^{-1}\|y-f(a)\|$ we obtain by applying $(d_a f)^{-1}$

$$((d_a f)^{-1}(y-f(a)) - (f^{-1}(y)-a))/\|y-f(a)\| \to 0 \text{ as } y \to f(a).$$

Hence f^{-1} is differentiable at $f(a)$ with $d_{f(a)}f^{-1} = (d_a f)^{-1}$. For the same reasons f^{-1} is differentiable at every point $y \in f(U')$ with $d_y f^{-1} = (d_{f^{-1}(y)}f)^{-1}$. Thus f^{-1} is C^1. □

COROLLARY (IMPLICIT FUNCTION THEOREM). *Let $U \subseteq R^{m+n}$ be a definable open set and $f_1, \ldots, f_n : U \to R$ definable C^1-functions. Let $(x_0, y_0) \in U$ be such that $f_1(x_0,y_0) = \cdots = f_n(x_0,y_0) = 0$ and the $n \times n$ matrix $((\partial f_j/\partial y_k)(x_0,y_0))_{1\leq j,k\leq n}$ is invertible. Then there are open definable neighborhoods V of x_0 in R^m and W of y_0 in R^n respectively, and there is a definable C^1-map $\phi: V \to W$ such that $V \times W \subseteq U$, $\phi(x_0) = y_0$, and such that for all $(x,y) \in V \times W$ we have*

$$f_1(x,y) = \cdots = f_n(x,y) = 0 \Leftrightarrow y = \phi(x).$$

PROOF. Apply the inverse function theorem to the map

$$(x,y) \mapsto (x, f_1(x,y), \ldots, f_n(x,y)) : U \to R^{m+n}. \quad \square$$

(2.12) EXERCISES.

1. (L'Hôpital's rule) Let I be an interval and $f, g : I \to R$ definable functions, and let a be one of the endpoints of the interval, possibly $a = +\infty$ or $a = -\infty$. Suppose that $g'(x) \neq 0$ for all $x \in I$ in some neighborhood of a, and that $\lim_{x \to a} f(x) = \lim_{x \to a} g(x) = 0$, or $\lim_{x \to a} |f(x)| = \lim_{x \to a} |g(x)| = +\infty$. Then

$$\lim_{x \to a} (f(x)/g(x)) = \lim_{x \to a} (f'(x)/g'(x)).$$

(Note that both limits exist in R_∞, by Chapter 3, (1.6).)

2. (Taylor's formula) Suppose the definable function $f : I \to R$ is $(n+1)$ times differentiable on the interval I, and let $a, b \in I$, $a < b$. Then

$$f(b) = f(a) + f'(a)(b-a) + \frac{f^{(2)}(a)}{2!}(b-a)^2 + \cdots + \frac{f^{(n)}(a)}{n!}(b-a)^n + \frac{f^{(n+1)}(z)}{(n+1)!}(b-a)^{n+1}$$

for some z with $a < z < b$.

§3. Definable maps are piecewise C^1

(3.1) Another basic tool is a C^1-version of cell decomposition. First we extend the notion of C^1-map and define C^1-cells.

DEFINITION.

(1) A definable map $f : A \to R^n$, where $A \subseteq R^m$ is not necessarily open, is said to be a **C^1-map** if there are a definable open set $U \subseteq R^m$ containing A and a definable C^1-map $F : U \to R^n$ such that $F|A = f$. (Then f is continuous, and for open A this gives the usual notion of C^1-map on an open set.)

(2) The notion of **C^1-cell** is defined inductively as in Chapter 3, (2.3), except that when forming $\Gamma(f)$ and (f, g) we now require the functions f and g (when R-valued) to be C^1 (and definable) instead of just continuous (and definable).

Note that every inclusion map $A \to R^m$ (for definable $A \subseteq R^m$) is C^1, and that if $f : A \to R^n$ is C^1 and $g : B \to R^p$ is C^1, with $B \subseteq R^n$, then $g \circ f : f^{-1}(B) \to R^p$ is C^1. Also $f = (f_1, \ldots, f_n) : A \to R^n$ is C^1 iff each function $f_i : A \to R$ is C^1.

(3.2) Theorem (C^1-cell decomposition).

(I_m) *For any definable sets $A_1, \ldots, A_k \subseteq R^m$ there is a decomposition of R^m into C^1-cells partitioning $A_1, \ldots, A_k$.*

(II_m) *For every definable function $f : A \to R$, $A \subseteq R^m$, there is a decomposition of R^m into C^1-cells, partitioning A, such that each restriction $f|C : C \to R$ is C^1 for each cell $C \subseteq A$ of the decomposition.*

If f and A are as in (II_m) and p is an interior point of A, we define

$$\nabla f(p) := ((\partial f/\partial x_1)(p), \ldots, (\partial f/\partial x_m)(p)),$$

provided these partials exist at p. If some partial is not defined at p, then ∇f is not defined at p. Further we put

$$A' := \{p \in A : p \text{ is an interior point of } A \text{ at which } \nabla f \text{ is defined}\}.$$

Along with (I_m) and (II_m) we will prove the following technical result.

(III_m) *$A - A'$ has empty interior.*

Proof of the theorem. By induction on m: (I_1) is trivial and (III_1) is a reformulation of (2.5). To prove (II_1), let $f : A \to R$, with $A \subseteq R$, be definable. By (2.5) there is a decomposition $\mathcal{D}$ of R partitioning A such that the restriction of f to each interval in $\mathcal{D}$ contained in A is differentiable, and by the monotonicity theorem we may assume, after refining $\mathcal{D}$ if necessary, that the restriction of f to each interval in $\mathcal{D}$ has continuous derivative.

Now we shall assume inductively that (I_d), (II_d) and (III_d) hold for all $d \leq m$, and derive successively (I_{m+1}), (III_{m+1}) and (II_{m+1}).

Proof of (I_{m+1}). Let $A_1, \ldots, A_k \subseteq R^{m+1}$ be definable. We want to find a C^1-decomposition of R^{m+1} partitioning $A_1, \ldots, A_k$. By ordinary cell decomposition there is a decomposition $\mathcal{D}$ of R^{m+1} partitioning $A_1, \ldots, A_k$. Then $\pi(\mathcal{D})$ is a decomposition of R^m where $\pi : R^{m+1} \to R^m$ is the usual projection map. Let $\pi(\mathcal{D}) = \{C_1, \ldots, C_n\}$, and for each $i = 1, \ldots, n$ let the cells of $\mathcal{D}$ that project onto C_i be

$$(-\infty, f_{i1}),\ \Gamma(f_{i1}),\ (f_{i1}, f_{i2}), \ldots,\ \Gamma(f_{is}),\ (f_{is}, +\infty),$$

where $f_{i1}, \ldots, f_{is} : C_i \to R$ are definable and continuous, $s = s(i)$. By (I_m) and (II_m) we may assume, after suitably refining $\pi(\mathcal{D})$, and $\mathcal{D}$ accordingly, that all C_i and all f_{ij} are C^1. Then $\mathcal{D}$ is a C^1-decomposition as required. □

Proof of (III_{m+1}). Let $f : A \to R$ be definable, $A \subseteq R^{m+1}$, and define A' as above. Consider an open box $U \times (a, b) \subseteq A$. It suffices to show that $U \times (a, b)$ intersects A'. By (2.5) and definable choice we can pick for each $p \in U$ an interval

$(\alpha(p), \beta(p)) \subseteq (a, b)$ such that α and β are definable R-valued functions on U and $\partial f/\partial x_{m+1}$ is defined on $\{p\} \times (\alpha(p), \beta(p))$. Using cell decomposition we can shrink U so that α and β are continuous on U; shrinking U further and changing a and b we may as well assume that $\partial f/\partial x_{m+1}$ is defined on the entire box $U \times (a, b)$. Take any $t \in (a, b)$. By applying (III_m) to the function $p \mapsto f(p, t) : U \to R$ we see that there must exist $p_0 \in U$ such that all partials $(\partial f/\partial x_i)(p_0, t)$ for $i = 1, \ldots, m$ are defined. Hence $(p_0, t) \in A'$, as desired. □

PROOF OF (II_{m+1}). Let $f : A \to R$ be definable, $A \subseteq R^{m+1}$. Let A' be as above. Take a decomposition $\mathcal{D}$ of R^{m+1} partitioning A and A' such that ∇f (and hence f) is continuous on each open cell of $\mathcal{D}$ contained in A'.

CLAIM. For each cell $C \in \mathcal{D}$ with $C \subseteq A$, $\dim(C) < m+1$, there is a decomposition $\mathcal{D}_C$ of R^{m+1} partitioning C such that $f|D$ is C^1 for each $D \in \mathcal{D}_C$.

PROOF OF CLAIM. Let $C \in \mathcal{D}$, $C \subseteq A$, $\dim(C) = d \leq m$, and consider the homeomorphism $p_C : C \to p(C)$ onto the open cell $p(C) \subseteq R^d$. By (II_d) we can partition $p(C)$ into finitely many cells B such that $f \circ p_C^{-1}|B$ is C^1.

By composing with the C^1-map p_C we obtain that $f|p_C^{-1}(B)$ is C^1. Take for $\mathcal{D}_C$ a decomposition that partitions each of the sets $p_C^{-1}(B)$. This proves the claim. □

Now we are done: by (I_{m+1}) there is a decomposition $\mathcal{D}'$ that refines $\mathcal{D}$ and all $\mathcal{D}_C$ as in the claim, and that consists moreover entirely of C^1-cells. Let $C' \in \mathcal{D}'$ and $C' \subseteq A$. It suffices to show that $f|C'$ is C^1. Take the cell $C \in \mathcal{D}$ that contains C', so $C \subseteq A$. If C is open, then by (III_{m+1}) the cell C intersects A', hence $C \subseteq A'$, so $f|C$ is C^1, so $f|C'$ is C^1. If $C \in \mathcal{D}$ is not open, apply the claim to conclude that $f|C'$ is C^1. □

(3.3) EXERCISES.

Let $f = (f_1, \ldots, f_n) : U \to R^n$ be a definable map on a (definable) open set $U \subseteq R^m$. We define inductively what it means for f to be a C^k-map, where k is a positive integer. For $k = 1$ this has been defined in (2.6). For $k > 1$ the map f is said to be a C^k-map if f is a C^1-map and $df : U \to R^{nm}$ is a C^{k-1}-map.

Next, let $f : A \to R^n$ be a definable map where $A \subseteq R^m$ is not necessarily open. Then we define f to be a C^k-map by replacing in (3.1) everywhere "C^1" by "C^k". Similarly, define C^k-cells as in (3.1) by replacing everywhere "C^1" by "C^k".

1. Show that the remarks at the end of (3.1) go through with "C^1" replaced by "C^k".

2. State and prove the C^k-Cell Decomposition Theorem, $k \geq 1$.

§4. Existence of good directions

(4.1) For each $x \in R^m$ with $\|x\| < 1$, put

$$v(x) := \left(x_1, \ldots, x_m, \sqrt{1 - \|x\|^2}\right),$$

so that $\|v(x)\| = 1$, $v(x) \in R^{m+1}$. (Note: $v(x)$ is the point on the unit sphere $S^m \subseteq R^{m+1}$ lying directly above x.) Let $A \subseteq R^{m+1}$ be definable with $\dim(A) < m + 1$. We now prove the final result of this chapter, the existence of "good directions" for A. It will be crucial in triangulating definable sets in the next chapter.

(4.2) THEOREM (GOOD DIRECTIONS LEMMA). *Let $B \subseteq R^m$ be a box contained in the disc $\|x\| < 1$. Then there is $x \in B$ such that for each point $p \in R^{m+1}$ the set $\{t \in R : p + t \cdot v(x) \in A\}$ is finite.*

For such x, every line in the affine space R^{m+1} with direction $v(x)$ intersects A in only finitely many points. We then call $v(x)$ a **good** direction. So the theorem tells us that the set of good directions is dense in the unit sphere $S^m \subseteq R^{m+1}$.

PROOF. Suppose there is no such x. Then for each $x \in B$ there is $p \in A$ and $\epsilon > 0$ such that $p + t \cdot v(x) \in A$ for all t with $|t| < \epsilon$. By definable choice there is then a definable map assigning to each $x \in B$ a pair (p, ϵ) with these properties. Applying (3.2) we may even assume, after shrinking B, that there are a C^1-map $P: B \to R^{m+1}$ and a fixed $\epsilon > 0$ such that $P(x) + t \cdot v(x) \in A$, for all $x \in B$ and $|t| < \epsilon$. Define the C^1-map $\phi: B \times R \to R^{m+1}$ by: $\phi(x,t) := P(x) + t \cdot v(x)$. Let us fix a point $x \in B$.

CLAIM. There is t, $|t| < \epsilon$, such that the R-linear map $d_{(x,t)}\phi : R^{m+1} \to R^{m+1}$ is invertible.

To prove the claim we consider the matrix $M(t)$ of $d_{(x,t)}\phi$ with respect to the standard basis $e_1, \ldots, e_m, e_{m+1}$. (In this context the elements of R^{m+1} are column vectors written as $(a_1, \ldots, a_{m+1})^t$, the "t" standing for "transpose".) Let $P = (P_1, \ldots, P_{m+1})$ and put $p_i := \left((\partial P_1/\partial x_i)(x), \ldots, (\partial P_{m+1}/\partial x_i)(x)\right)^t$ for $1 \leq i \leq m$. A straightforward computation shows that the columns of $M(t)$ are given by

$$\begin{aligned} M(t) \cdot e_i &= p_i + t \cdot \left(e_i - \left(x_i/\sqrt{1 - \|x\|^2}\right) \cdot e_{m+1}\right), \ 1 \leq i \leq m, \\ M(t) \cdot e_{m+1} &= v(x). \end{aligned}$$

It follows that $\det\big(M(t)\big)$ is a polynomial in t. For $t \neq 0$ the $m+1$ columns of $M(t)$ are independent if and only if the $m + 1$ vectors

$$(p_i/t) + e_i - \left(x_i/\sqrt{1 - \|x\|^2}\right) \cdot e_{m+1} \ (i = 1, \ldots, m), \text{ and } v(x)$$

are independent. When t is large these $m+1$ vectors are close to the $m+1$ vectors

$$e_i - \left(x_i/\sqrt{1-\|x\|^2}\right)\cdot e_{m+1}\ (i=1,\ldots,m), \text{ and } v(x),$$

which are easily seen to be independent. So for sufficiently large t the $m+1$ columns of $M(t)$ are independent. Therefore the polynomial $\det(M(t))$ is not identically zero. So $\det(M(t)) \neq 0$ for some t, $|t| < \epsilon$, which proves the claim.

Take such a t, $|t| < \epsilon$, with $d_{(x,t)}\phi$ invertible. From (2.11) it follows that ϕ is injective on a box B_0 around (x,t). Take B_0 so small that $B_0 \subseteq B \times (-\epsilon, \epsilon)$. Then $\phi(B_0) \subseteq A$ by definition of ϕ. But $\dim(\phi(B_0)) = \dim(B_0) = m+1$, contradicting $\dim(A) < m+1$. □

(4.3) EXERCISES.

1. Let $A \subseteq R^{m+1}$ be a definable set of dimension $\leq m$. Call a unit vector $u \in S^m \subseteq R^{m+1}$ an **asymptotic direction for** A if for each $\epsilon > 0$ and $r > 0$, there is a point $x \in A$ with $\|x\| > r$ and $\|(x/\|x\|) - u\| < \epsilon$. Show that the (definable) set of asymptotic directions for A is of dimension $< m$. (In particular, not every unit vector is an asymptotic direction for A.)

2. Let $A \subseteq R^m$ be definable and $\dim(A) \leq k < m$. Show that there is an $(m-k)$-dimensional linear subspace L of R^m all of whose translates $v + L$ ($v \in R^m$) meet A in only finitely many points. (Hint: proceed by induction on $m-k$.)

3. In this exercise, assume that R is an ordered field, not necessarily real closed.

Let $A \subseteq R^m$ be semilinear with $\dim(A) < m$. Show that A is contained in a finite union of affine subspaces of R^m of dimension $< m$, and derive that there is a good direction for A, that is, a nonzero vector $u \in R^m$ such that each line $p + Ru$ ($p \in R^m$) intersects A in only finitely many points.

Notes and comments

Proposition (2.5) that definable functions are piecewise differentiable is adapted from the appendix in [19], which also shows C^1-cell decomposition in the classical real case assuming "strong o-minimality". The proof of the inverse function theorem is along the lines of that for polynomial maps in Brumfiel [7]. The good directions lemma (4.2) was inspired by a lemma due to Koopman and Brown [36], which also appeared in (unpublished?) lecture notes by Sussmann on subanalytic sets (mid 1980s).

CHAPTER 8

TRIANGULATION

Introduction

We work in this chapter with a fixed o-minimal expansion $(R, <, \mathcal{S})$ of an ordered real closed field $(R, <, 0, 1, +, -, \cdot)$.

We point out, however, that most of Section 1 is of a purely semilinear character, and makes sense over any ordered field. In Section 2 we prove the main result in this chapter, the triangulation theorem, which allows us to reduce many questions to the semilinear case. In Section 3 we use triangulation in this way to show that a definable continuous function $f: A \to R$ on a definable closed subset A of a definable set B can be continuously and definably extended to B. This result is needed in the next two chapters on trivialization and quotient spaces.

§1. Simplexes and complexes

For simplicity, we denote the ordered field $(R, <, 0, 1, +, -, \cdot)$ just by R. In this section the role of R is mainly that of an ordered vector space over itself. In particular we will only deal with subsets of R^n that are semilinear over R, in the sense of Chapter 1, (7.9). We call a map $f: A \to R^n$ with $A \subseteq R^m$ **semilinear** if its graph $\Gamma(f) \subseteq R^{m+n}$ is semilinear.

(1.1) An **affine** subspace of R^n of dimension d is by definition a translate $L + a$ of a linear subspace L of R^n of dimension d. For $d = 1$ this is also called a **line**, and for $d = 2$ we speak of a **plane**. Given points $a_0, a_1, \ldots, a_k$ in R^n we have

$$\begin{aligned} a_0 + R(a_1 - a_0) + \cdots + R(a_k - a_0) &= \left\{\sum t_i a_i : t_i \in R, \sum t_i = 1\right\} \\ &= \text{the smallest affine subspace containing } a_0, \ldots, a_k. \end{aligned}$$

This affine subspace is called the **affine span** of $a_0, \ldots, a_k$.

(1.2) A tuple $a_0, \ldots, a_k$ of points in R^n is called **affine independent** if the affine span of $a_0, \ldots, a_k$ has dimension k. For $k = 0$ this is always the case, for $k = 1$ it means that $a_0 \neq a_1$, for $k = 2$ that a_0, a_1, a_2 are distinct and not on a line.

Affine independence of $a_0, a_1, \ldots, a_k$ is equivalent to the condition that the k vectors $a_1 - a_0, \ldots, a_k - a_0$ are linearly independent.

Let $a_0, \ldots, a_k$ *be affine independent,* $a_i \in R^n$. Then the map $(t_0, \ldots, t_k) \mapsto \sum t_i a_i$ is a homeomorphism from the affine subspace $\{(t_0, \ldots, t_k) \in R^{k+1} : \sum t_i = 1\}$ of R^{k+1} onto the affine span of $a_0, \ldots, a_k$. In particular, each point x in the affine span of $a_0, \ldots, a_k$ has a *unique* representation as

$$x = t_0 a_0 + \cdots + t_k a_k \quad \text{with} \sum t_i = 1,$$

and $t_0, \ldots, t_k$ are called the **affine coordinates** of the point x with respect to $a_0, \ldots, a_k$. (Sometimes also called **barycentric** coordinates.)

(1.3) A set $S \subseteq R^n$ is called **convex** if for every two distinct points a, b in S the line segment $[a, b] := \{ta + (1-t)b : 0 \le t \le 1\}$ is contained in S. The **convex hull** of a set $S \subseteq R^n$ is the smallest convex subset of R^n that contains S, and is easily seen to equal the set of all points $\sum t_i a_i$ with $a_0, \ldots, a_k \in S$ and $\sum t_i = 1$, $t_i \ge 0$.

Given a convex set C in R^n, a point $p \in C$ is called an **extreme** point of C if for all $p_1, p_2 \in C$ such that $p = (p_1 + p_2)/2$ we have $p = p_1 = p_2$.

(1.4) Definitions. An affine independent tuple of points $a_0, a_1, \ldots, a_k \in R^n$ is said to **span the simplex**

$$(a_0, \ldots, a_k) := \left\{\sum t_i a_i : \text{all } t_i > 0, \sum t_i = 1\right\} \subseteq R^n.$$

We call $(a_0, \ldots, a_k)$ a k**-simplex** in R^n. Note that it is *open* in the affine span of $a_0, \ldots, a_k$. Its dimension as a semilinear set is easily seen to be k.

The (topological) closure of $(a_0, \ldots, a_k)$ in R^n is denoted by $[a_0, \ldots, a_k]$, so that

$$\begin{aligned}[a_0, \ldots, a_k] &:= \left\{\sum t_i a_i : \text{all } t_i \ge 0, \sum t_i = 1\right\} \\ &= \text{the convex hull of } \{a_0, \ldots, a_k\}.\end{aligned}$$

One easily checks that $a_0, \ldots, a_k$ are exactly the extreme points of $[a_0, \ldots, a_k]$. We call $a_0, \ldots, a_k$ the **vertices** of $(a_0, \ldots, a_k)$ and also of $[a_0, \ldots, a_k]$. Instead of "S is the set of vertices of the simplex σ" we also say "S spans σ".

A **face** of $(a_0, \ldots, a_k)$ is a simplex spanned by a nonempty subset of $\{a_0, \ldots, a_k\}$. Note that distinct nonempty subsets of $\{a_0, \ldots, a_k\}$ span *disjoint* faces, and that $[a_0, \ldots, a_k]$ is the (disjoint) union of the faces of $(a_0, \ldots, a_k)$.

Given simplexes σ and τ we write $\sigma < \tau$ if σ is a **proper** face of τ, that is, σ is a face of τ and $\sigma \ne \tau$.

The **barycenter** $b(\sigma)$ of a k-simplex $\sigma = (a_0, \ldots, a_k)$ is the point

$$b(\sigma) = \frac{1}{k+1}(a_0 + \cdots + a_k),$$

which belongs to σ.

EXAMPLES OF CLOSURES OF SIMPLEXES.

$$\begin{aligned}
[a_0] &= (a_0) = \{a_0\}; \text{ however, } a_i \notin (a_0, \ldots, a_k) \text{ if } k > 0; \\
[a_0, a_1] &= (a_0, a_1) \cup \{a_0\} \cup \{a_1\}, \text{ the line segment between} \\
&\quad \text{distinct points } a_0 \text{ and } a_1; \\
[a_0, a_1, a_2] &= \text{the “triangle” spanned by distinct points } a_0, a_1, a_2 \text{ not on a line;} \\
[a_0, a_1, a_2, a_3] &= \text{the “tetrahedron” spanned by four distinct points } a_0, a_1, a_2, a_3 \\
&\quad \text{that do not all lie on one plane.}
\end{aligned}$$

(1.5) DEFINITIONS. A **complex in R^n** is a finite collection K of simplexes in R^n, such that for all $\sigma_1, \sigma_2 \in K$,

$$\begin{aligned}
\text{either } \operatorname{cl}(\sigma_1) \cap \operatorname{cl}(\sigma_2) &= \emptyset, \\
\text{or } \operatorname{cl}(\sigma_1) \cap \operatorname{cl}(\sigma_2) &= \operatorname{cl}(\tau)
\end{aligned}$$

for some common face τ of σ_1 and σ_2. (This τ is not required to belong to K, which makes our definition somewhat more general than is usual. The usual definition of "complex" corresponds to what we call "closed complex" below.)

Note that distinct simplexes in K are disjoint. We put

$$\begin{aligned}
|K| &:= \text{ union of the simplexes of } K, \text{ a bounded semilinear subset of } R^n; \\
\operatorname{Vert}(K) &:= \text{ the set of vertices of the simplexes in } K, \text{ a finite subset of } R^n.
\end{aligned}$$

We also call $|K|$ the **polyhedron** spanned by the complex K. Note: the notational conflict that $|K|$ also denotes the number of simplexes of K is always resolved in favor of "$|K|$ = polyhedron spanned by K". (All bounded semilinear sets in R^n are polyhedrons, that is, of the form $|K|$ for a suitable complex K in R^n, see (2.14), exercise 2.)

A subset of a complex K in R^n is also a complex in R^n, and is called a **subcomplex of K**.

A complex K is called **closed** if it contains with each simplex all its faces. Equivalently, a closed complex in R^n is a finite collection K of disjoint simplexes in R^n such that each face of a simplex in K also belongs to K. Note also that a complex K in R^n is closed iff $|K|$ is closed in R^n.

Given a complex K in R^n we put $\overline{K}$:= the set of faces of the simplexes in K. So $\overline{K}$ is clearly a closed complex in R^n, with K as a subcomplex, and

$$\operatorname{cl}|K| = |\overline{K}|.$$

Let L be a subcomplex of K. Then $|L|$ is closed in $|K|$ if and only if $\overline{L} \cap K = L$. In that case we also say that L is **closed in** K.

Note that a complex K in R^n is completely determined by the finite set $\text{Vert}(K) \subseteq R^n$ and by the collection of subsets of $\text{Vert}(K)$ that span simplexes in K.

(1.6) Let K and L be complexes in R^m and R^n respectively.

Then a **vertex map** $V: K \to L$ is a map $V: \text{Vert}(K) \to \text{Vert}(L)$ such that whenever $(a_0, \ldots, a_k)$ is a k-simplex of K, then $\{V(a_0), \ldots, V(a_k)\}$ spans a simplex of L. (We allow $V(a_i) = V(a_j)$ for $i \neq j$.)

Such a vertex map $V: K \to L$ determines a continuous map $|V|: |K| \to |L|$ by:

$$|V|\left(\sum t_i a_i\right) = \sum t_i V(a_i)$$

where $(a_0, \ldots, a_k)$ is a k-simplex of K, $t_i > 0$, $\sum t_i = 1$.

To show that $|V|$ is continuous, show first that the restriction of $|V|$ to the closure of $(a_0, \ldots, a_k)$ in $|K|$ is continuous. Then use the fact that if $f: X \to Y$ is a map between topological spaces X and Y, and $X = X_1 \cup \cdots \cup X_k$ a covering of X by finitely many closed sets X_i such that each restriction $f|X_i: X_i \to Y_i$ is continuous, then f is continuous. This general fact will be used frequently.

Note also that the map $|V|$ is semilinear, and that if $W: L \to M$ is a second vertex map, then the composition $W \circ V: K \to M$ is a vertex map such that

$$|W \circ V| = |W| \circ |V|.$$

We now state our main result concerning triangulation.

(1.7) Triangulation theorem. *Each definable set $S \subseteq R^m$ is definably homeomorphic to a polyhedron $|K|$ for some complex K in R^m.*

This theorem expresses the truly remarkable fact that the (definable) topology of a definable set can be completely described in *finite combinatorial terms.* To see this, let K be a complex, and consider its *scheme*

$$\left(\text{Vert}(K), \{S \subseteq \text{Vert}(K) : \ S \text{ spans a simplex of } K\}\right),$$

which is just a finite set equipped with a collection of subsets, a finite combinatorial object *par excellence.* For L to be a complex with isomorphic scheme means there is a bijective vertex map $V: K \to L$ whose inverse is also a vertex map. Such a map V induces a semilinear homeomorphism $|V|: |K| \to |L|$. In this sense, the scheme of K, as a finite combinatorial object, determines the topology of the polyhedron $|K|$ up to semilinear homeomorphism.

We prove this theorem in a more precise version in the next section. In the rest of this section we consider some elementary operations on complexes.

(1.8) BARYCENTRIC SUBDIVISION. Let K be a complex. A **K-flag** is a sequence

$$\mathfrak{F}: \sigma_0 < \sigma_1 < \cdots < \sigma_f \quad (f \geq 0)$$

of simplexes, with $\sigma_f \in K$ and each σ_i a proper face of the σ_j with $j > i$.

We do not require $\sigma_i \in K$ for $i < f$, or that σ_i is of dimension i.

To such a K-flag $\mathfrak{F}$ we associate an f-simplex $b(\mathfrak{F}) := \big(b(\sigma_0), \ldots, b(\sigma_f)\big)$ whose vertices are the barycenters of the simplexes of $\mathfrak{F}$. One easily checks that $b(\mathfrak{F})$ is indeed an f-simplex, that $b(\mathfrak{F}) \subseteq \sigma_f$, and that if $\mathfrak{F}_1$ and $\mathfrak{F}_2$ are distinct K-flags then $b(\mathfrak{F}_1)$ and $b(\mathfrak{F}_2)$ are disjoint.

The (first) **barycentric subdivision of K** is the complex K' whose simplexes are the simplexes $b(\mathfrak{F})$ associated to the K-flags $\mathfrak{F}$. One checks easily that the $b(\mathfrak{F})$ indeed form a complex, and that $|K'| = |K|$.

(1.9) We need one further general construction.

Let $(a_0, \ldots, a_k)$ be a k-simplex in R^n, and $r_i, s_i \in R$, $r_i \leq s_i$ for $i = 0, \ldots, k$, and $r_j < s_j$ for some j. Put $b_i := (a_i, r_i)$, $c_i := (a_i, s_i) \in R^{n+1}$. Then $(b_0, \ldots, b_k)$, $(c_0, \ldots, c_k)$ are k-simplexes in R^{n+1}. One also easily checks that if $b_i \neq c_i$ (that is, $r_i < s_i$) then $(b_0, \ldots, b_i, c_i, \ldots, c_k)$ is a $(k+1)$-simplex. We now construct a complex L in R^{n+1} such that $|L|$ is the set of points "between $[b_0, \ldots, b_k]$ and $[c_0, \ldots, c_k]$".

(1.10) LEMMA. *Let L consist of all $(k+1)$-simplexes $(b_0, \ldots, b_i, c_i, \ldots, c_k)$ with $b_i \neq c_i$, and all faces of these $(k+1)$-simplexes. Then L is a closed complex and*

$$\begin{aligned} |L| &= \Big\{ t\big(t_0 b_0 + \cdots + t_k b_k\big) + (1-t)\big(t_0 c_0 + \cdots + t_k c_k\big) : \\ &\qquad 0 \leq t \leq 1,\ t_i \geq 0,\ \sum t_i = 1 \Big\} \\ &= \textit{convex hull of } \big\{ b_0, \ldots, b_k, c_0, \ldots, c_k \big\}. \end{aligned}$$

PROOF. The proof of this lemma is rather long, and is divided into four parts.

PART I. In this part we show L *is a closed complex.* Let σ and τ be faces of $(b_0, \ldots, b_i, c_i, \ldots, c_k)$ and $(b_0, \ldots, b_j, c_j, \ldots, c_k)$ respectively, $r_i < s_i$, $r_j < s_j$.

Suppose $\sigma \cap \tau \neq \emptyset$. We have to show that $\sigma = \tau$. This is clear if $i = j$ since then σ and τ are faces of the same simplex. So assume $i < j$, and let $x \in \sigma \cap \tau$. Write

$$\begin{aligned} x &= t_0 b_0 + \cdots + t_i b_i + t c_i + t_{i+1} c_{i+1} + \cdots + t_k c_k, \text{ with } t + t_0 + \cdots + t_k = 1, \\ x &= u_0 b_0 + \cdots + u_{j-1} b_{j-1} + u b_j + u_j c_j + \cdots + u_k c_k, \text{ with } u + u_0 + \cdots + u_k = 1. \end{aligned}$$

Applying the projection map $\pi : R^{n+1} \to R^n$ to these expressions for x and considering the affine coordinates of $\pi(x)$ with respect to $a_0, \ldots, a_k$ we get

$$(*) \quad t_i + t = u_i,\ t_j = u_j + u, \text{ and } t_\lambda = u_\lambda \text{ for } \lambda \in \{0, \ldots, k\},\ \lambda \neq i,\ \lambda \neq j.$$

Also the $(n+1)^{\text{th}}$ coordinate of x satisfies

$$\begin{aligned} x_{n+1} &= t_0 r_0 + \cdots + t_i r_i + t s_i + t_{i+1} s_{i+1} + \cdots + t_k s_k \\ &= u_0 r_0 + \cdots + u_{j-1} r_{j-1} + u r_j + u_j s_j + \cdots + u_k s_k. \end{aligned}$$

Subtracting $t_0 r_0 + \cdots + t_{i-1} r_{i-1} + t_{j+1} s_{j+1} + \cdots + t_k s_k$ from both sums yields

$$t_i r_i + t s_i + \sum_{i+1 \leq \lambda \leq j} t_\lambda s_\lambda = \sum_{i \leq \lambda \leq j-1} u_\lambda r_\lambda + u r_j + u_j s_j.$$

Subtracting the right side from the left and using $(*)$ we get

$$t(s_i - r_i) + \sum_{i+1 \leq \lambda \leq j-1} t_\lambda (s_\lambda - r_\lambda) + u(s_j - r_j) = 0,$$

in which each term on the left is nonnegative. Hence each term equals zero, so $t = u = 0$, and $t_\lambda = 0$ whenever $i+1 \leq \lambda \leq j-1$ and $r_\lambda < s_\lambda$. The vertices of σ are exactly the points among $b_0, \ldots, b_i, c_i, \ldots, c_k$ with respect to which x has a nonzero affine coordinate, and similarly with τ and $b_0, \ldots, b_j, c_j, \ldots, c_k$. This shows that σ and τ have the same vertices, so that $\sigma = \tau$.

PART II. In this part we prove that $|L|$ contains the set

$$\left\{ t\big(t_0 b_0 + \cdots + t_k b_k\big) + (1-t)\big(t_0 c_0 + \cdots + t_k c_k\big) : 0 \leq t \leq 1,\ t_i \geq 0,\ \sum t_i = 1 \right\}.$$

Let $x = t\big(t_0 b_0 + \cdots + t_k b_k\big) + (1-t)\big(t_0 c_0 + \cdots + t_k c_k\big)$, $0 \leq t \leq 1, t_i \geq 0$, $\sum t_i = 1$. We have to show that $x \in |L|$. Clearly the $(n+1)^{\text{th}}$ coordinate x_{n+1} of x satisfies

$$t_0 r_0 + \cdots + t_k r_k \leq x_{n+1} \leq t_0 s_0 + \cdots + t_k s_k,$$

so there is an $i \in \{0, \ldots, k\}$ such that $r_i < s_i$ and

$$\begin{aligned} &t_0 r_0 + \cdots + t_{i-1} r_{i-1} + t_i r_i + t_{i+1} s_{i+1} + \cdots + t_k s_k \leq x_{n+1} \\ &\leq t_0 r_0 + \cdots + t_{i-1} r_{i-1} + t_i s_i + t_{i+1} s_{i+1} + \cdots + t_k s_k. \end{aligned}$$

Hence $x_{n+1} = t_0 r_0 + \cdots + t_{i-1} r_{i-1} + a + t_{i+1} s_{i+1} + \cdots + t_k s_k$, with $t_i r_i \leq a \leq t_i s_i$. Then we can write $a = u r_i + v s_i$ with $u, v \geq 0$ and $u + v = t_i$. Thus

$$x_{n+1} = t_0 r_0 + \cdots + t_{i-1} r_{i-1} + u r_i + v s_i + t_{i+1} s_{i+1} + \cdots + t_k s_k.$$

Also

$$\pi(x) = \sum t_i a_i = t_0 a_0 + \cdots + t_{i-1} a_{i-1} + u a_i + v a_i + t_{i+1} a_{i+1} + \cdots + t_k a_k.$$

Therefore

$$x = t_0 b_0 + \cdots + t_{i-1} b_{i-1} + u b_i + v c_i + t_{i+1} c_{i+1} + \cdots + t_k c_k,$$

since the right hand side has the same $(n+1)^{\text{th}}$ coordinate and the same projection in R^n as x. This expression for x shows that $x \in [b_0, \ldots, b_i, c_i, \ldots, c_k]$, and finishes the proof of part II.

PART III. The set $S := \{t \sum t_i b_i + (1-t) \sum t_i c_i : 0 \leq t \leq 1,\ t_i \geq 0,\ \sum t_i = 1\}$ is convex.

To prove this, first a preliminary remark.

For a point $x \in S$, $x = t \sum t_i b_i + (1-t) \sum t_i c_i$ with $0 \leq t \leq 1$, $t_i \geq 0$, $\sum t_i = 1$, we have $\pi(x) = \sum t_i a_i \in [a_0, \ldots, a_k]$. Hence, given a point $p = \sum t_i a_i \in [a_0, \ldots, a_k]$, $t_i \geq 0$, $\sum t_i = 1$, the vertical line $\pi^{-1}\{p\} = (p, 0) + Re_{n+1}$ intersects S exactly in the line segment $[\sum t_i b_i, \sum t_i c_i]$, and the set of $(n+1)^{\text{th}}$ coordinates of the points on this line segment equals $[\sum t_i r_i, \sum t_i s_i]$.

Let now $x, y \in S$, and consider a convex combination $z = ux + vy$, $u, v \geq 0$, $u + v = 1$. We have to show that $z \in S$. Write $x = t \sum t_i b_i + (1-t) \sum t_i c_i$ as above, and similarly $y = t' \sum t'_i b_i + (1-t') \sum t'_i c_i$. Then

$$\pi(z) = u\pi(x) + v\pi(y) = \sum (ut_i + vt'_i) a_i \in [a_0, \ldots, a_k],$$

and the $(n+1)^{\text{th}}$ coordinates satisfy

$$z_{n+1} = ux_{n+1} + vy_{n+1}, \quad \sum t_i r_i \leq x_{n+1} \leq \sum t_i s_i$$

and $\sum t'_i r_i \leq y_{n+1} \leq \sum t'_i s_i$. Hence $\sum (ut_i + vt'_i) r_i \leq z_{n+1} \leq \sum (ut_i + vt'_i) s_i$, so that by the preliminary remark we have $z \in S$.

PART IV. Rest of the proof. The set S from part III obviously contains the points $b_0, \ldots, b_k, c_0, \ldots, c_k$, hence it contains the convex hull of these points. Finally, the convex hull of these points contains each set $[b_0, \ldots, b_i, c_i, \ldots, c_k]$ with $b_i \neq c_i$, hence it contains the union $|L|$ of these sets. This completes the circle of inclusions, and thereby the proof of the lemma. □

(1.11) REMARKS. One should note that the complex L of our lemma depends on the listing of the vertices of $(a_0, \ldots, a_k)$ in the order indicated: first a_0, then a_1, etc. On the other hand, its polyhedron $|L|$ is independent of this ordering. Note also that the simplexes $(b_0, \ldots, b_k)$ and $(c_0, \ldots, c_k)$ are in L, and that $\pi(\sigma)$ is a face of $(a_0, \ldots, a_k)$ for each $\sigma \in L$.

(1.12) Exercises.

To state the first exercise, define a map $f : R^m \to R^n$ to be **affine** if it is of the form $f(x) = L(x) + a$ where $L : R^m \to R^n$ is linear, and $a \in R^n$. Note that for $n = 1$ this is just what we called an affine function in Chapter 1, Section 7.

1. Show that the composition of affine maps is affine, and that if $f : R^m \to R^n$ is affine, then $f(\sum t_i a_i) = \sum t_i f(a_i)$ for $a_i \in R^m$, and $t_i \in R$ with $\sum t_i = 1$. Conclude that if two affine maps $R^m \to R^n$ take the same value at points $a_0, \ldots, a_k \in R^m$, they agree on the affine subspace of R^m spanned by $a_0, \ldots, a_k$.

2. Let $T : R^m \to R^m$ be an R-linear automorphism, or a translation, and let $a_0, \ldots, a_k$ be an affine independent tuple of points in R^m. Show that the tuple $T(a_0), \ldots, T(a_k)$ is affine independent and that

$$T\big((a_0, \ldots, a_k)\big) = \big(T(a_0), \ldots, T(a_k)\big).$$

3. Let σ be a k-simplex in R^m. Show that there are $m-k$ polynomials $f_1, \ldots, f_{m-k} \in R[X_1, \ldots, X_m]$ and $k+1$ polynomials $g_0, \ldots, g_k \in R[X_1, \ldots, X_m]$, with all f's and g's of degree 1, such that

$$\sigma = \big\{x \in R^m : f_i(x) = 0,\ g_j(x) > 0,\ 1 \le i \le m-k,\ 0 \le j \le k\big\},$$

and

$$\mathrm{cl}(\sigma) = \big\{x \in R^m : f_i(x) = 0,\ g_j(x) \ge 0,\ 1 \le i \le m-k,\ 0 \le j \le k\big\}.$$

Hint: let the affine space spanned by the vertices of σ be given by the system of equations $f_i(x) = 0$ $(1 \le i \le m-k)$.

4. Let $V : K \to L$ be a vertex map between complexes K and L, and let $\sigma \in K$. Show that $|V|(\sigma)$ is the simplex of L spanned by $\big\{V(p) : p \text{ is a vertex of } \sigma\big\}$.

5. Let K be a complex in R^m, $\sigma \in K$ and $\dim(\sigma) = \dim\big(|K|\big)$. Show that then σ is open in $|K|$.

6. (Carathéodory) Show that the convex hull of a set $S \subseteq R^m$ is the union of the convex hulls of its finite subsets of size $\le m+1$. (Hint: assume $|S| = m+2$.)

7. Show that the convex hull of a definable set in R^m is definable.

(1.13) Remark. Everything in this section (except (1.7) and exercise 7 above) makes sense and is valid for an arbitrary ordered field R, not necessarily real closed.

§2. Triangulation theorem

First some facts on extending continuous definable functions.

(2.1) LEMMA. *Let σ be a k-simplex in R^m, $k > 0$, and $f : \partial\sigma \to R$ a continuous definable function. Then f has a continuous definable extension $g : \mathrm{cl}(\sigma) \to R$.*

PROOF. Let $b = b(\sigma)$ be the barycenter of σ. Connect each point $p \in \partial\sigma$ to b by the line segment $[p, b] = \{tp + (1-t)b : 0 \le t \le 1\}$, and define g on this line segment by $g(tp + (1-t)b) = tf(p)$. This gives g the value 0 at the barycenter b, and g is well defined on all of $\mathrm{cl}(\sigma)$ since the line segments cover $\mathrm{cl}(\sigma)$ and line segments corresponding to different p's only intersect at the point b. Moreover g extends f. By viewing $\mathrm{cl}(\sigma)$ as the continuous image of the map

$$(t, p) \mapsto tp + (1-t)b \ : \ [0,1] \times (\mathrm{cl}(\sigma) - \sigma) \to R^m$$

we can appeal to Chapter 6, (1.13) to conclude that g is continuous. □

(2.2) LEMMA. *Let K be a closed complex in R^m and L a closed subcomplex. Then each continuous definable function $f : |L| \to R$ has a continous definable extension $\tilde{f} : |K| \to R$.*

PROOF. It will suffice to show that if $L \neq K$, then we can find a strictly larger closed subcomplex L' of K and a continuous definable extension $f' : |L'| \to R$. Assuming $L \neq K$, take a simplex $\sigma \in K - L$ of minimal dimension. If σ is a 0-simplex, then $\sigma = \{a\}$ for a point $a \notin |L|$, $L \cup \{\sigma\}$ is a strictly larger closed subcomplex of K, and f has continuous definable extension $f' : |L \cup \{\sigma\}| = |L| \cup \{a\} \to R$ given by $f'(a) = 0$. Next assume σ is a k-simplex with $k > 0$. Then all proper faces of σ are in L, so $\mathrm{cl}(\sigma) - \sigma \subseteq |L|$, and by the previous lemma the function $f|\,\mathrm{cl}(\sigma) - \sigma$ extends continuously to a definable function $g : \mathrm{cl}(\sigma) \to R$. Further $L' := L \cup \{\sigma\}$ is a strictly larger closed subcomplex, and f extends continuously to the function $f' : |L'| = |L| \cup \mathrm{cl}(\sigma) \to R$ defined by

$$\begin{aligned} f'(x) &= f(x) \text{ for } x \in |L|, \\ f'(x) &= g(x) \text{ for } x \in \mathrm{cl}(\sigma). \end{aligned}$$ □

(2.3) DEFINITION. Let $A \subseteq R^m$ be a definable set. A **triangulation in R^n** of A is a pair (Φ, K) consisting of a complex K in R^n and a definable homeomorphism $\Phi : A \to |K|$. Note that then

$$\Phi^{-1}(K) := \{\Phi^{-1}(\sigma) : \sigma \in K\}$$

is a finite partition of A. We call $(A, \Phi^{-1}(K))$ a **triangulated set**. The triangulation is said to be **compatible with the subset** $A' \subseteq A$ if A' is a union of elements of $\Phi^{-1}(K)$. Note that by Chapter 6. (1.10), we have

A is closed and bounded $\Leftrightarrow$ the complex K is closed.

In that case each continuous definable R-valued function on $\text{cl}(C)$, where $C \in \Phi^{-1}(K)$, has a continuous definable R-valued extension to A, by lemma (2.2).

(2.4) Let $(A, \mathcal{P})$ be a triangulated set, and $C, D \in \mathcal{P}$.

We call D a **face** of C if $D \subseteq \text{cl}(C)$ and a **proper face** of C if $D \subseteq \text{cl}(C) - C$.

Note that if $\mathcal{P} = \Phi^{-1}(K)$ for the triangulation (Φ, K) of A, then D is a (proper) face of C iff the simplex $\Phi(D)$ is a (proper) face of the simplex $\Phi(C)$. Hence,

If D is a proper face of C and $y \in D$, then there are arbitrarily small definable neighborhoods U of y in A such that $U \cap C$ is definably connected. (To see this, use the fact that a convex definable set is definably connected.)

Note also that $\text{cl}(C) \cap A =$ union of the faces of C in $\mathcal{P}$, for $C \in \mathcal{P}$.

(2.5) DEFINITION. **A multivalued function F on the triangulated set** $(A, \mathcal{P})$ is a finite collection of functions, $F = \{f_{C,i} : C \in \mathcal{P},\ 1 \leq i \leq k(C)\}$, $k(C) \geq 0$, each $f_{C,i} : C \to R$ continuous and definable, and $f_{C,1} < \cdots < f_{C,k(C)}$. We set

$$\begin{aligned} \Gamma(F) &:= \bigcup_{f \in F} \Gamma(f), \\ F|C &:= \{f_{C,i} : 1 \leq i \leq k(C)\} \text{ for } C \in \mathcal{P}, \\ \mathcal{P}^F &:= \{\Gamma(f) : f \in F\} \cup \{(f_{C,i}, f_{C,i+1}) : C \in \mathcal{P},\ 1 \leq i < k(C)\}, \\ A^F &:= \text{the union of the sets in } \mathcal{P}^F, \end{aligned}$$

so $\mathcal{P}^F$ is a partition of $A^F \subseteq R^{m+1}$ and $\pi(\Gamma(F)) = \pi(A^F) \subseteq A$. Here $\pi : R^{m+1} \to R^m$ is the projection map onto the first m coordinates.

We consider two properties of a multivalued function F on $(A, \mathcal{P})$:

(1) We call F **closed** if for each pair $C, D \in \mathcal{P}$ with D a proper face of C and each $f \in F|C$ there is $g \in F|D$ such that $g(y) = \lim_{x \to y} f(x)$ for all $y \in D$.

Note that then each $f \in F$, say $f \in F|C$, extends continuously to a definable function $\text{cl}(f) : \text{cl}(C) \cap A \to R$, that the restrictions of $\text{cl}(f)$ to the faces of C in $\mathcal{P}$ belong to F, and that $\Gamma(F)$ is closed in $A \times R$. (Lemma (2.6) below is a converse to this.)

(2) We call F **full** if F is closed, $k(C) \geq 1$ for all $C \in \mathcal{P}$, and for each pair $C, D \in \mathcal{P}$ with D a proper face of C and each $g \in F|D$ we have $g = \text{cl}(f)|D$ for some $f \in F|C$, where $\text{cl}(f)$ is the continuous extension of f to $\text{cl}(C) \cap A$.

Note that then $\pi\Gamma(F) = A$, and that if A is closed and bounded, so is A^F.

(2.6) LEMMA. *Let F be a multivalued function on the triangulated set $(A, \mathcal{P})$ such that $\Gamma(F)$ is closed in $A \times R$, and there is $M > 0$ such that $\Gamma(F) \subseteq A \times [-M, M]$. Then F is closed.*

PROOF. Let $C, D \in \mathcal{P}$, D a proper face of C, and let $f \in F|C$. Given $y \in D$ we have to show first that $\lim_{x \to y} f(x)$ exists. Set $l := \liminf_{x \to y} f(x)$, $L := \limsup_{x \to y} f(x)$; i.e. $l = \inf\{t \in R : \text{there are } x \in C \text{ arbitrarily close to } y \text{ with } f(x) < t\}$, and similarly for L. Clearly $-M \leq l \leq L \leq +M$, and $l = L$ implies that $\lim_{x \to y} f(x)$ exists and equals l. Suppose $l < L$; take any t with $l < t < L$. By the *remark* in (2.4) there are arbitrarily small definable neighborhoods U of y in A such that $U \cap C$ is definably connected. Since f takes both values $< t$ and values $> t$ on $U \cap C$ it takes also the value t on $U \cap C$. Hence $(y, t) \in \mathrm{cl}(\Gamma(f)) \subseteq \Gamma(F)$ for each t in the interval (l, L). But there can only be finitely many t with $(y, t) \in \Gamma(F)$. This contradiction shows $l = L = \lim_{x \to y} f(x)$. Set $\phi(y) = \lim_{x \to y} f(x)$. One checks easily that $\phi : D \to R$ is continuous, and since $\Gamma(F)$ is closed it follows that $\phi \in F$. □

(2.7) MORE TERMINOLOGY. Let (Φ, K) be a triangulation in R^n of $A \subseteq R^m$, and let $B \subseteq R^{m+1}$ be a definable set, $B \subseteq A \times R$. Then a triangulation (Ψ, L) in R^{n+1} of B is said to be a **lifting** of (Φ, K) if $K = \{\pi(\sigma) : \sigma \in L\}$ and the diagram

$$\begin{array}{ccc} B & \xrightarrow{\ \Psi\ } & |L| \\ \downarrow & & \downarrow \\ A & \xrightarrow[\ \Phi\ ]{} & |K| \end{array}$$

commutes where the vertical maps are restrictions of the projection maps $R^{m+1} \to R^m$ onto the first m coordinates and $R^{n+1} \to R^n$ onto the first n coordinates. (In particular, B projects onto A.)

The triangulation (Φ, K) of $A \subseteq R^m$ gives rise to the triangulation (Φ, K') of A, where K' is as in (1.8). We call (Φ, K') the **barycentric subdivision of** (Φ, K).

Here is the main triangulation lemma.

(2.8) LEMMA. *Let (Φ, K) be a triangulation in R^n of the closed and bounded definable set $A \subseteq R^m$ and let F be a full multivalued function on the triangulated set $(A, \Phi^{-1}(K))$. Then (Φ, K) or its barycentric subdivision (Φ, K') can be lifted to a triangulation (Ψ, L) in R^{n+1} of A^F that is compatible with the sets in $\Phi^{-1}(K)^F$.*

PROOF. Note first that K is closed. Replacing K if necessary by its barycentric subdivision K' and F by the corresponding set of restrictions, we may assume in addition to fullness of F that the following condition holds for each $C \in \Phi^{-1}(K)$:

$$(*) \qquad \begin{cases} \text{If } f_1, f_2 \in F|C,\ f_1 \neq f_2, \text{ then there is at least one vertex of} \\ C \text{ where } \mathrm{cl}(f_1) \text{ and } \mathrm{cl}(f_2) \text{ take different values.} \end{cases}$$

Here the vertices of C are $\Phi^{-1}(a_0), \ldots, \Phi^{-1}(a_k)$ with $a_0, \ldots, a_k$ the vertices of $\Phi(C)$.

Fix a linear order on $\mathrm{Vert}(K)$. We now construct L and Ψ above each $C \in \Phi^{-1}(K)$. Let $a_0, \ldots, a_k$ be the vertices of $\Phi(C)$ listed in the order we imposed on $\mathrm{Vert}(K)$.

Let $f < g$ be two *successive* members of $F|C$. Put

$$r_i := \mathrm{cl}(f)(\Phi^{-1}(a_i)),\ s_i := \mathrm{cl}(g)(\Phi^{-1}(a_i)),\ b_i := (a_i, r_i),\ c_i := (a_i, s_i) \in R^{n+1}.$$

Because of $(*)$ the conditions for lemma (1.10) are satisfied. Let $L(f,g)$ be the complex in R^{n+1} constructed in that lemma, so $|L(f,g)|$ is the convex hull of $\{b_0, \dots, b_k, c_0, \dots, c_k\}$. Define the map $\Psi_{f,g}: [\mathrm{cl}(f), \mathrm{cl}(g)] \to |L(f,g)|$ by

$$\Psi_{f,g}(x, t\,\mathrm{cl}(f)(x) + (1-t)\,\mathrm{cl}(g)(x)) = t\Phi_b(x) + (1-t)\Phi_c(x),\ 0 \le t \le 1,$$

where $\Phi_b(x)$, $\Phi_c(x)$ are the points of $[b_0, \dots, b_k]$ and $[c_0, \dots, c_k]$ with the same affine coordinates with respect to $b_0, \dots, b_k$ and $c_0, \dots, c_k$ as $\Phi(x) \in [a_0, \dots, a_k]$ has with respect to $a_0, \dots, a_k$. Clearly $\Psi_{f,g}$ is a bijection, and it follows from Chapter 6, (1.13)(ii) that $\Psi_{f,g}$ is continuous.

We also define for each $f \in F|C$ the closed complex $L(f)$ in R^{n+1}: it consists of the k-simplex $(b_0, \dots, b_k)$ with $b_i = (a_i, \mathrm{cl}(f)(\Phi^{-1}(a_i)))$ as above, and all its faces. Then $\Psi_f: \Gamma(\mathrm{cl}(f)) \to |L(f)|$ is by definition the homeomorphism given by

$$\Psi_f(x, \mathrm{cl}(f)(x)) = \Phi_b(x)$$

where $\Phi_b(x)$ is defined as before.

Now let L be the union of all complexes $L(f,g)$ and $L(f)$, for all $C \in \Phi^{-1}(K)$. A lengthy but routine argument using fullness of F shows L is a closed complex in R^{n+1}, and that any two among the $\Psi_{f,g}$'s and Ψ_f's coincide on the intersection of their domains. (In the interest of conciseness we leave here the numerous details to the reader.) Hence these maps glue to a continuous definable map $\Psi: A^F \to |L|$. One verifies easily that Ψ is a bijection. Hence, by Chapter 6, (1.12), Ψ is homeomorphism. Using the last remark of (1.11) it is routine to check that the triangulation (Ψ, L) lifts (Φ, K). Finally, note that (Ψ, L) is compatible with the sets in $\Phi^{-1}(K)^F$. □

(2.9) TRIANGULATION THEOREM. *Let $S \subseteq R^m$ be a definable set, with definable subsets $S_1, \dots, S_k$. Then S has a triangulation in R^m that is compatible with these subsets.*

PROOF. By induction on m, the case $m = 0$ being trivial. Suppose the triangulation theorem holds for a certain value of m. To prove that each definable set $S \subseteq R^{m+1}$ has a triangulation in R^{m+1} compatible with any given definable subsets $S_1, \dots, S_k$, we first reduce to the case that S is closed and bounded. Let $\mu: R \to (-1,1)$ be the definable homeomorphism $t \mapsto t/\sqrt{1+t^2}$, and S^μ and S_i^μ the images of S and S_i under the homeomorphism $(x_1, \dots, x_m) \mapsto (\mu(x_1), \dots, \mu(x_m)): R^m \to (-1,1)^m$. It suffices to prove that the closure of S^μ, a closed and bounded set in R^{m+1}, has a triangulation in R^{m+1} that is compatible with its subsets S^μ and $S_1^\mu, \dots, S_k^\mu$. So, adjusting our notations, we may as well assume from the start that S is already closed and bounded. Under this assumption on S, let

$$T := \mathrm{bd}(S) \cup \mathrm{bd}(S_1) \cup \dots \cup \mathrm{bd}(S_k),$$

so $\dim(T) < m + 1$ by Chapter 4, (1.10). Note also that T is closed and bounded.

By the good directions lemma (4.2) in Chapter 7 the set T has a good direction vector. Choose a linear automorphism of R^{m+1} that maps this good direction vector to the vertical unit vector e_{m+1}. To obtain a triangulation of S in R^{m+1} compatible with $S_1, \ldots, S_k$ we may replace S and the S_i's by their images under this linear automorphism, which replaces T also by its image, so we may assume

(*) the unit vector e_{m+1} is a good direction for the set T.

Under this assumption we shall prove that there is a triangulation of $\pi(S) \subseteq R^m$ in R^m that can be lifted to a triangulation of S compatible with $S_1, \ldots, S_k$.
Set $A := \pi(S) = \pi(T)$, a closed and bounded subset of R^m. By (*) and cell decomposition the set T is the disjoint union of $\Gamma(f)$'s for finitely many definable continuous functions f on cells A_j that form a partition of A. By the inductive hypothesis there is a triangulation (Φ, K) of A in R^m that is compatible with each set A_j. The nonempty restrictions of the f's to the sets of $\Phi^{-1}(K)$ form a multivalued function F on $\big(A, \Phi^{-1}(K)\big)$ such that $\Gamma(F) = T$. Since T is closed, lemma (2.6) implies that F is closed. Unfortunately, F may not be full.

Therefore we modify F as follows to $\tilde{F}$. Each $f \in F$ extends, first continuously to the closure of its domain, and then, by the last remark of (2.3), further to a continuous definable function $\tilde{f}: A \to R$. Change T to $\tilde{T} :=$ union of the graphs $\Gamma(\tilde{f})$ for $f \in F$, and take a new triangulation $(\tilde{\Phi}, \tilde{K})$ of A in R^m that is compatible with all sets in $\Phi^{-1}(K)$, all sets $\pi\big(S \cap \Gamma(\tilde{f})\big)$ and $\pi\big(S_i \cap \Gamma(\tilde{f})\big)$ for $f \in F$, $i \in \{1, \ldots, k\}$, and such that the nonempty restrictions of the $\tilde{f}$'s (for $f \in F$) to the sets of $\mathcal{P} := \tilde{\Phi}^{-1}(\tilde{K})$ form a multivalued function $\tilde{F}$ on $(A, \mathcal{P})$. Clearly $\Gamma(\tilde{F}) = \tilde{T}$. Since $\tilde{T}$ is closed, $\tilde{F}$ is closed by lemma (2.6), and therefore $\tilde{F}$ is full. Hence by the last lemma we can lift $(\tilde{\Phi}, \tilde{K})$ or its barycentric subdivision to a triangulation (Ψ, L_0) of $A^{\tilde{F}}$ that is compatible with the sets of $\mathcal{P}^{\tilde{F}}$. From the compatibility of $(\tilde{\Phi}, \tilde{K})$ with the sets mentioned it follows that if $B \in \mathcal{P}$ and $g \in \tilde{F}|B$ then $\Gamma(g) \subseteq S$, or $\Gamma(g) \cap S = \emptyset$, and similarly for S_i, $i \in \{1, \ldots, k\}$. Also, if $B \in \mathcal{P}$ and g, h are successive members of $\tilde{F}|B$ it follows from Chapter 6, (3.3), that $(g, h) \subseteq S$ or $(g, h) \cap S = \emptyset$, and similarly for each S_i, $1 \leq i \leq k$. Hence $\mathcal{P}^{\tilde{F}}$ partitions S and each S_i. Let $L := \{\sigma \in L_0 : \sigma \subseteq \Psi(S)\}$. Then $(\Psi|S, L)$ is clearly a triangulation of S compatible with $S_1, \ldots, S_k$ and it lifts $(\tilde{\Phi}, \tilde{K})$ or its barycentric subdivision. □

(2.10) The following example shows decisively that the use of a linear automorphism (alias a linear change of coordinates) in the proof above cannot be omitted. Something like the choice of a "good direction" seems essential for triangulation.

EXAMPLE. Let $S := \big\{(x, y, z) \in R^3 : y = xz,\ |x|, |z| \leq 1\big\}$. Then, with $\pi: R^3 \to R^2$ the usual projection map we have $\pi(S) = \big\{(x, y) \in R^2 : |y| \leq |x| \leq 1\big\}$. We claim

No triangulation of $\pi(S)$ can be lifted to a triangulation of S.

To see this, consider a triangulation (Φ, K) of $\pi(S)$, and suppose this triangulation can be lifted to a triangulation $(\hat{\Phi}, \hat{K})$ of S. We shall derive a contradiction. Let $\mathcal{P} := \Phi^{-1}(K)$ and $\hat{\mathcal{P}} := \hat{\Phi}^{-1}(\hat{K})$, so $\mathcal{P} = \{\pi(A) : A \in \hat{\mathcal{P}}\}$. Let $T := \{(0,0,z) : |z| \leq 1\} \subseteq S$. Then T is a union of sets in $\hat{\mathcal{P}}$ since $(0,0)$ must be a vertex of $\mathcal{P}$ (exercise). In particular, T contains a one-dimensional set $A \in \hat{\mathcal{P}}$, corresponding to a 1-simplex under $\hat{\Phi}$. Let $a, b \in T$ be the two vertices of A.

Since A is not open in S, it is a proper face of a set $B \in \hat{\mathcal{P}}$ corresponding to a 2-simplex under $\hat{\Phi}$ (exercise). Then $\dim(B - T) = 2$, and π is injective on $S - T$, so $\dim(\pi(B)) = 2$. Hence $\pi(B) \in \mathcal{P}$ corresponds to a 2-simplex under Φ, and the vertices of $\pi(B)$ are the images under π of the vertices of B. But π collapses the distinct vertices a, b of B to the single point $(0,0)$ in $\pi(B)$, so $\pi(B)$ cannot have three distinct vertices, contradiction.

(2.11) DIGRESSION 1: CRITERION FOR DEFINABLE EQUIVALENCE.

Call definable sets $A \subseteq R^m$ and $B \subseteq R^n$ **definably equivalent** if there is a definable bijection from A onto B. Clearly "definable equivalence" is an equivalence relation on the class of definable sets.

COROLLARY. *Let $A \subseteq R^m$ and $B \subseteq R^n$ be definable sets. Then*

A and B are definably equivalent $\Leftrightarrow$ $\dim(A) = \dim(B)$ *and* $E(A) = E(B)$.

PROOF. The $\Rightarrow$-direction follows from Chapter 4. Suppose now that A and B have the same dimension and the same Euler characteristic. By our triangulation theorem we may also assume that A and B are finite disjoint unions of simplexes in R^m and R^n respectively. For this case we prove a bit more:

CLAIM. A and B are semilinearly equivalent, that is, there is a semilinear bijection from A onto B.

Before starting the proof of this claim, first three general observations.

(1) A k-simplex is, for $k > 0$, the disjoint union of two k-simplexes and a $(k-1)$-simplex. (Exercise, induction on k.)
(2) Any two k-simplexes are semilinearly homeomorphic. (By remark in (1.2).)
(3) $E(\sigma) = (-1)^k$ for a k-simplex σ. (See exercise 1 below.)

PROOF OF CLAIM. By induction on $k = \dim(A) = \dim(B)$. For $k = 0$ the sets A and B are finite of the same cardinality $E(A) = E(B)$, and any bijection will do. Let $k > 0$. Write A and B as finite disjoint unions of simplexes:

$$A = \sigma_1 \cup \cdots \cup \sigma_p \cup \cdots \cup \sigma_q,$$
$$B = \tau_1 \cup \cdots \cup \tau_r \cup \cdots \cup \tau_s,$$

where the σ_i and τ_j are of maximal dimension k for $1 \le i \le p$ and $1 \le j \le r$, and of lower dimension for $i > p$ and $j > r$.

Using observation (1) we may arrange that $p = r$, and that there is at least one $(k-1)$-simplex among the σ_i for $i > p$, as well as at least one $(k-1)$-simplex among the τ_j for $j > p$. By observation (2) there is a semilinear bijection between $\sigma_1 \cup \cdots \cup \sigma_p$ and $\tau_1 \cup \cdots \cup \tau_p$. Also $A' := \sigma_{p+1} \cup \cdots \cup \sigma_q$ and $B' := \tau_{p+1} \cup \cdots \cup \tau_s$ have both dimension $k-1$, and the same Euler characteristic, by observation (3). So by the inductive hypothesis, A' and B' are semilinearly equivalent. Hence A and B are semilinearly equivalent. □

(2.12) DIGRESSION 2: NUMBER OF DEFINABLE HOMEOMORPHISM TYPES.

Let us say that definable sets $A \subseteq R^m$ and $B \subseteq R^n$ are **definably homeomorphic** if there is a definable homeomorphism $f : A \to B$. This defines an equivalence relation on the collection of definable sets in R^m, $m = 0, 1, 2, \ldots$, and a **definable homeomorphism type** is simply an equivalence class for this equivalence relation. An easy consequence of the triangulation theorem is that *there are only countably many definable homeomorphism types*. To see this, suppose that a (nonempty) definable set $A \subseteq R^m$ is definably homeomorphic to $|K|$ where K is a complex with N vertices. Let $e_1, \ldots, e_N$ be the basis vectors of R^N.

CLAIM. A is definably homeomorphic to a union of faces of the simplex $(e_1, \ldots, e_N)$.

Just take a bijection $V : \text{Vert}(K) \to \{e_1, \ldots, e_N\}$, and let L be the complex with $\text{Vert}(L) = \{e_1, \ldots, e_N\}$ such that $V : K \to L$ is a vertex map whose inverse is also a vertex map; then $|V| : |K| \to |L|$ is a definable homeomorphism, and the claim is an obvious consequence.

(2.13) We close this section with some exercises. Exercise 3 is intended for readers familiar with elementary model theory. Moreover, exercise 4 depends on the result of exercise 3. Actually, in the next chapter we obtain essentially the same results as stated in the exercises without any model theory. I include the exercise nevertheless, since it illustrates a typical model-theoretic method.

(2.14) EXERCISES.

1. Let Δ_m be the m-simplex in R^m spanned by $0, e_1, \ldots, e_m$, where $e_1, \ldots, e_m$ are the standard basis vectors. Show that

$$\Delta_{m+1} = \left\{(x, r) \in R^{m+1} : \ x \in \Delta_m, \ 0 < r < 1 - \sum x_i\right\}.$$

Derive from this that Δ_m is an open cell in R^m, and that each m-simplex in R^m is definably homeomorphic to an open cell in R^m.

2. Show that each bounded semilinear set is a polyhedron. More generally, let $S_1, \ldots, S_k$ be semilinear subsets of a bounded semilinear set $S \subseteq R^m$. Show that

there is a complex K in R^m such that $S = |K|$, and each S_i is a union of simplexes of K. (The result of this exercise holds for any ordered field R.)

3. Let $S \subseteq R^{m+n}$ be definable. Show that the sets S_a fall into only finitely many definable homeomorphism types as the parameter a ranges over R^m. More precisely, show there is a definable map $f: S \to R^N$ (for some N) such that for each $a \in R^m$ the map $f_a: S_a \to R^N$ is a homeomorphism from S_a onto a union of faces of $(e_1, \ldots, e_N)$.

4. Let $S \subseteq R^{m+n}$ be definable. Show that there are a partition of R^m into definable sets $A_1, \ldots, A_k$ and for each $i = 1, \ldots, k$ a definable set $F_i \subseteq R^n$ and a definable homeomorphism $h_i : S \cap (A_i \times R^n) \xrightarrow{\sim} A_i \times F_i$ such that the diagram

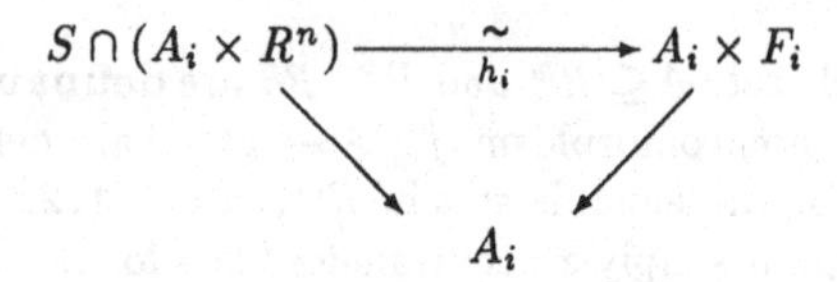

commutes. (Here the downward arrows indicate the obvious projection maps.) Hint: use the result of exercise 3 in combination with Chapter 6, (2.5), exercise 2.

§3. Definable retractions and definable continuous extensions

As one might expect, the triangulation theorem has numerous consequences, and we develop a few here that are needed in later chapters.

(3.1) Homotopies, retractions and extensions.

Let $A \subseteq R^m$ and $B \subseteq R^n$ be definable sets, and $f, g: A \to B$ definable continuous maps. A **definable homotopy between f and g** is a definable continuous map $H: A \times [0,1] \to B$ such that $f(x) = H(x,0)$ and $g(x) = H(x,1)$ for all x in A. (It may help to view H as a "continuous" family of maps $(H_t)_{0 \le t \le 1}$, with $H_t: A \to B$ given by $H_t(x) = H(x,t)$, so $f = H_0$ and $g = H_1$.)

A definable set A is called **definably contractible to the point** $a \in A$ if there is a definable homotopy $H: A \times [0,1] \to A$ between the identity map on A and the map $A \to A$ taking the constant value a. (Such an H is called a **definable contraction from A to the point** a.)

Note that for a point $a \in R^m$ the map $H: R^m \times [0,1] \to R^m$ given by

$$H(x,t) = (1-t)x + ta$$

is a definable contraction from R^m to a. Since each cell is definably homeomorphic to some R^m it follows that each cell is definably contractible to each of its points.

Let a definable subset A' of a definable set $A \subseteq R^m$ be given, and denote by $i_{A'}$ the inclusion map $A' \to A$.

A map $r : A \to A'$ is called a **definable retraction** if r is a definable continuous map such that $r(x) = x$ for all $x \in A'$. So a definable retraction $r : A \to A'$ satisfies $r \circ i_{A'} = 1_{A'}$.

Next we define a **definable contraction from** A **to** A' to be a definable homotopy $H : A \times [0,1] \to A$ between 1_A and $i_{A'} \circ r$, for a definable retraction $r : A \to A'$. This map r is then uniquely determined by $r(x) = H(x,1)$, and if we want to be more explicit, we call H a **definable contraction from** A **to** r. If in addition $H(a', t) = a'$ for all $a' \in A'$ and $t \in [0,1]$ we call H a **definable strong deformation retraction from** A **to** A'. Note that for $a \in A$ a definable contraction from A to the point a is the same as a definable contraction from A to the subset $\{a\}$.

We also need one important construction in complexes:

(3.2) Let K be a complex.

Given a definable set $A \subseteq |K|$ we define the **star** $st_K(A)$ of A in K to be the union of all simplexes $\sigma \in K$ such that $\text{cl}(\sigma) \cap A \neq \emptyset$. Clearly $st_K(A)$ is an open neighborhood of A in $|K|$, and in fact the smallest neighborhood of A in $|K|$ that is a union of simplexes of K.

We can now formulate a basic result on the existence of nice retractions.

(3.3) PROPOSITION. *Let K be a complex and L a subcomplex of K, closed in K. Then there is a definable retraction $r : st_{K'}(|L|) \to |L|$ such that for each point $x \in st_{K'}(|L|) - |L|$ the open line interval $(x, r(x))$ lies entirely in the simplex of K' that contains the point x.*

(3.4) REMARKS.

(1) We shall give an explicit definition of the retraction r.

(2) Given such r the map $H : st_{K'}(|L|) \times [0,1] \to st_{K'}(|L|)$ given by

$$H(x,t) = (1-t) \cdot x + t \cdot r(x)$$

is a definable strong deformation retraction from $st_{K'}(|L|)$ to $|L|$. The image $r(S)$ of each simplex $S \subseteq st_{K'}(|L|)$ of K' is contained in $\text{cl}(S)$.

(3.5) Before proving (3.3) we need some preparations. Let $E := \text{Vert}(K) = \text{Vert}(\overline{K})$. For each vertex $e \in E$ we define a function $\lambda_e : |\overline{K}| \to [0,1]$ by setting for a point $x \in (e_0, \ldots, e_k)$, for a k-simplex $(e_0, \ldots, e_k) \in \overline{K}$,

$$\begin{aligned} \lambda_e(x) &:= 0 \text{ if } e \notin \{e_0, \ldots, e_k\}, \\ &:= \lambda_i \text{ if } e = e_i, \text{ where } x = \sum \lambda_j e_j,\ \sum \lambda_j = 1, \\ &\qquad \text{with } 0 < \lambda_j \text{ for } j \in \{0, \ldots, k\}. \end{aligned}$$

We call λ_e the **affine coordinate function of $\overline{K}$ corresponding to the vertex** e. Note that λ_e is definable and continuous, and for each $x \in |\overline{K}|$ we have

$$x = \sum_{e \in E} \lambda_e(x) \cdot e,$$

$$1 = \sum_{e \in E} \lambda_e(x).$$

Now $\{b(\sigma) : \sigma \in \overline{K}\}$ is the set of vertices of the first barycentric subdivision $\overline{K}'$ of $\overline{K}$, and so we also have for each $\sigma \in \overline{K}$ the affine coordinate function $\lambda_{b(\sigma)}$ of $\overline{K}'$ corresponding to the vertex $b(\sigma)$. For convenience we denote $\lambda_{b(\sigma)}$ by λ_σ, so $\lambda_\sigma : |\overline{K}'| = |\overline{K}| \to [0,1]$ is a definable (continuous) function.

Next we introduce for $\sigma \in \overline{K}$ a "weight function" $w_\sigma : |\overline{K}| \to [0,1]$ as follows:

(i) if $\sigma \in L$, then w_σ is identically 1,
(ii) if $\sigma \in \overline{K} - \overline{L}$, then w_σ is identically 0,
(iii) if $\sigma \in \overline{L} - L$, then

$$w_\sigma(x) = \frac{\sum_{\sigma < \tau \in L} \lambda_\tau(x)}{\sum_{\sigma < \tau \in K} \lambda_\tau(x)},$$

provided the denominator $\sum_{\sigma < \tau \in K} \lambda_\tau(x)$ is nonzero, and $w_\sigma(x) = 0$ otherwise.

(The summations are over the $\tau \in L$ with $\sigma < \tau$, respectively over the $\tau \in K$ with $\sigma < \tau$.) Clearly w_σ is definable, though it may not be continuous.

Now define $\mu_\sigma : |K| \to [0,1]$ by $\mu_\sigma(x) := w_\sigma(x)\lambda_\sigma(x)$. Clearly μ_σ is definable, and by the following lemma μ_σ is also continuous.

(3.6) LEMMA. *Let $\sigma \in \overline{L} - L$, and let $S \in K'$ be a simplex on which the function $\sum_{\sigma < \tau \in K} \lambda_\tau$ vanishes. Then λ_σ also vanishes on S.*

PROOF. Let $S = (b(\sigma_0), \ldots, b(\sigma_n))$ with $\sigma_0 < \cdots < \sigma_n$ a K-flag, $\sigma_n \in K$. Suppose λ_σ is nonzero at some point of S, hence at every point of S. Then σ is among $\sigma_0, \ldots, \sigma_{n-1}, \sigma_n$, but $\sigma = \sigma_n$ would imply $\sigma \in K \cap \overline{L} = L$, contradiction. Hence σ is among $\sigma_0, \ldots, \sigma_{n-1}$, in particular, $n \geq 1$. Thus $\sigma < \sigma_n \in K$, so that

$$\sum_{\sigma < \tau \in K} \lambda_\tau(x) > 0,$$

for all $x \in S$. $\square$

(3.7) Lemma. *For each $x \in st_{K'}(|L|)$ there is $\sigma \in L$ with $\mu_\sigma(x) > 0$.*

Proof. Let $S \in K'$ be contained in $st_{K'}(|L|)$. As above, $S = (b(\sigma_0), \ldots, b(\sigma_n))$ for some K-flag $\sigma_0 < \cdots < \sigma_n$, $\sigma_n \in K$. Some face of S intersects $|L|$, hence $\sigma_k \in L$ for some $k \in \{0, \ldots, n\}$. Then we have for $\sigma = \sigma_k$: $\mu_\sigma(x) = \lambda_\sigma(x) > 0$ for all $x \in S$. □

We continue the proof of proposition (3.3). Suppose the complex K lies in R^m. Then we define the (definable, continuous) map $r : st_{K'}(|L|) \to R^m$ by

$$(*) \qquad r(x) = \frac{\sum \mu_\sigma(x) \cdot b(\sigma)}{\sum \mu_\sigma(x)}, \text{ with } \sigma \text{ ranging over } \overline{K}.$$

(3.8) Lemma. *$r(S) \subseteq \mathrm{cl}(S) \cap |L|$ for each $S \in K'$ contained in $st_{K'}(|L|)$, and $r(x) = x$ for $x \in |L|$.*

Thus $r(st_{K'}(|L|)) \subseteq |L|$ and for $x \in st_{K'}(|L|) - |L|$ the open line interval $(x, r(x))$ lies entirely in the simplex $S \in K'$ that contains the point x. So this lemma implies that r is a retraction with the properties stated in proposition (3.3).

Proof of the lemma. As in the previous lemma, let $S = (b(\sigma_0), \ldots, b(\sigma_n))$ with $\sigma_0 < \cdots < \sigma_n$, $\sigma_n \in K$. Take $k \in \{0, \ldots, n\}$ maximal with $\sigma_k \in L$. Consider a $\sigma \in \overline{K}$ such that μ_σ does not vanish identically on S. Then λ_σ does not vanish on S, so $\sigma = \sigma_t$ for some t. If $t > k$ the function w_σ, and hence also μ_σ would vanish identically on S. Thus $t \leq k$. This gives for each $x \in S$ a convex combination $r(x) = \sum_{0 \leq i \leq k} \alpha_i b(\sigma_i)$ with $0 < \alpha_i \in R$ for all i and $\sum \alpha_i = 1$, as is easily checked. Hence $r(S) \subseteq (b(\sigma_0), \ldots, b(\sigma_k)) \subseteq \mathrm{cl}(S) \cap |L|$. If $S \subseteq |L|$, then $\sigma_n \in L$, so $k = n$, and by distinguishing the cases $\sigma \in L$, $\sigma \in \overline{K} - \overline{L}$, and $\sigma \in \overline{L} - L$, we see that in each case $\mu_\sigma | S = \lambda_\sigma | S$. (For the case $\sigma \in \overline{L} - L$, use the fact that $K \cap \overline{L} = L$.) Thus $r(x) = x$ for all $x \in |L|$. □

This finishes the proof of proposition (3.3). In combination with triangulability it gives the following.

(3.9) Corollary. *Let A be a definable closed subset of the definable set $B \subseteq R^m$. Then there are a definable open subset U of B containing A, and a definable retraction $r : \mathrm{cl}(U) \cap B \to A$.*

Proof. By (3.3) there are a definable open neighborhood V of A in B and a definable retraction $r_V : V \to A$. Applying Chapter 6, (3.5) to the closed subsets A and $B - V$ of B we find definable U, open in B and containing A, such that $\mathrm{cl}(U) \cap B \subseteq V$. Now restrict r_V to $\mathrm{cl}(U) \cap B$. □

Here is an attractive further consequence, generalizing (2.2). It will be needed as a technical tool in the next chapter, on trivialization, to make a multivalued function full.

(3.10) COROLLARY. *Let A be a definable closed subset of the definable set $B \subseteq R^m$. Then each definable continuous function $f : A \to R$ can be extended to a definable continuous function $\tilde{f} : B \to R$. More generally, let $f : A \to C$ be a definable continuous map into a definable set $C \subseteq R^n$ that is definably contractible to a point $c \in C$. Then f can be extended to a definable continuous function $\tilde{f} : B \to C$.*

PROOF. Choose U and r as in (3.9). By Chapter 6, (3.8) there is a definable continuous function $\lambda : B \to [0,1]$ with $\lambda^{-1}(0) = A$ and $\lambda^{-1}(1) = B - U$. Finally, let Φ be a definable contraction from C to c. Then we define the extension $\tilde{f} : B \to C$ of f as follows:

$$\begin{aligned} \tilde{f}(x) &:= \Phi\big(f\big(r(x)\big), \lambda(x)\big), \text{ for } x \in \mathrm{cl}(U) \cap B, \\ &:= c, \text{ for } x \in B - U. \end{aligned}$$

One verifies easily that $\tilde{f}$ has the desired properties. □

Notes and comments

Section 2 of this chapter follows closely the proof of the semialgebraic triangulation theorem by Delfs and Knebusch in [11]. The main difference is that the good directions lemma is only a *partial* substitute for the use of Noether normalization in their proof. The way out of this difficulty is to first establish lemma (2.2) and use that to construct a full multivalued function as starting point for the inductive triangulation procedure. The idea of using multivalued functions and lemma (2.6) was suggested to me by reading some (unpublished?) lecture notes on subanalytic sets by H. Sussmann (mid 1980s).

The claim in (2.11) that two bounded semilinear sets are semilinearly equivalent if and only if they have the same dimension and the same Euler characteristic was made to me in conversation by Schanuel; see [51] for related results.

Section 3 of this chapter is adapted from the semialgebraic case treated by Delfs and Knebusch in [13], which contains further interesting material that extends easily to our setting.

See Dieudonné [16] for some history of triangulation results for manifolds and algebraic varieties, and Giesecke [27], Łojasiewicz [39] and Hironaka [31] for various kinds of triangulation theorems for semialgebraic and semianalytic sets.

The inductive way triangulation was established in this chapter provides extra information that has been used by Woerheide in [65] to prove the excision property for simplicial and singular homology for definable sets in the o-minimal setting. (This way of obtaining the excision property seems to be new also in the semialgebraic case. This case is extensively discussed by Knebusch in [34].)

CHAPTER 9

TRIVIALIZATION

Introduction

A continuous map $f : S \to A$ between topological spaces is often thought of as describing a "continuous" family of sets $\left(f^{-1}(a)\right)_{a\in A}$ parametrized by the space A. From this viewpoint the *trivial* maps are the simplest. One says that f is **trivial** if f looks like a projection map $A \times F \to A$, precisely, if there are a topological space F and a homeomorphism $h : S \to A \times F$ such that the following diagram commutes.

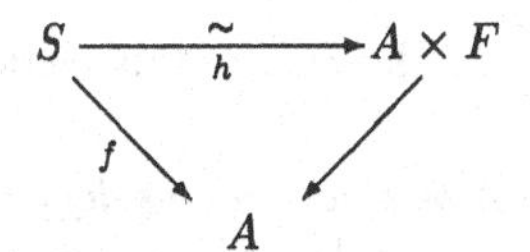

Note that then all fibers $f^{-1}(a)$ are homeomorphic to F, and that $h = (f, \lambda)$ for some unique continuous map $\lambda : S \to F$. We also say that f is trivial over a subspace B of A if the restriction $f|f^{-1}(B) : f^{-1}(B) \to B$ is trivial. (Then all fibers $f^{-1}(b)$ with $b \in B$ are mutually homeomorphic.)

As in the previous chapter we fix an o-minimal expansion $\mathcal{R} = (R, <, \mathcal{S})$ *of an ordered (necessarily real closed) field* $(R, <, 0, 1, +, -, \cdot)$.

The main result of the first section of this chapter generalizes a well-known theorem of Hardt [29] on semialgebraic continuous maps between real semialgebraic sets, and is roughly that if f, S, and A as above are definable—in the sense of $\mathcal{R}$ of course—then one can partition A into finitely many definable subsets over each of which f is trivial. This is even true in the stronger sense that the corresponding F's and h's can also be taken to be definable. This result was already established in Chapter 8, (2.14), exercise 4, but there it depended on a model-theoretic argument. The constructive proof given here is more elementary, though much longer. We also give some further refinements, which in Section 2 lead to applications like the conical local structure of definable sets. In Section 3 we show how Wilkie's theorem used in combination with our trivialization results implies new topological finiteness properties of polynomials with few terms. We solve in particular a problem of Risler and Benedetti on semialgebraic sets of bounded additive complexity. In this way we continue a theme initiated by Khovanskii [32, 33].

§1. Trivialization theorem

(1.1) Consider a definable map $f: S \to A$, where $A \subseteq R^m$ and $S \subseteq R^n$ are definable sets. Viewing A as a **base space** or parameter space, f describes the **family** of sets $\big(f^{-1}(a)\big)_{a \in A}$. A **definable trivialization** of f is a pair (F, λ) consisting of a definable set $F \subseteq R^N$, for some N, and a definable map $\lambda : S \to F$ such that $(f, \lambda): S \to A \times F$ is a homeomorphism.

So (f, λ) identifies S with the cartesian product $A \times F$, and under this identification f corresponds to the projection map $A \times F \to A$. Note that then f and λ are continuous, and that (f, λ) maps each fiber $f^{-1}(a)$ of f homeomorphically onto $\{a\} \times F$, in particular all fibers are definably homeomorphic to F.

We call f **definably trivial** if f has a **definable trivialization.** Given a definable subset $A' \subseteq A$ we call f **definably trivial over** A' if the restriction $f|f^{-1}(A') : f^{-1}(A') \to A'$ is definably trivial. Note that if f is definably trivial, then f is definably trivial over each definable subset of the base space A: if (F, λ) is a definable trivialization of f and $A' \subseteq A$ is definable, then $\big(F, \lambda|f^{-1}(A')\big)$ is a definable trivialization of $f|f^{-1}(A'): f^{-1}(A') \to A'$.

We can now formulate the main theorem of this section.

(1.2) THEOREM. *Let $f: S \to A$ be a continuous definable map as above. Then there is a finite partition $A = A_1 \cup \cdots \cup A_M$ of the base space A into definable sets A_i such that f is definably trivial over each A_i.*

We actually provide a stronger version where the definable trivializations over each A_i are also compatible with distinguished definable subsets of S.

First we establish an easy lemma on good directions. Next we prove a technical lemma involving homotopies and retractions, and then we can trivialize.

(1.3) Let $n > 0$, and $S^{n-1} := \{x \in R^n : \|x\| = 1\}$, the set of unit vectors. Recall that a unit vector $u \in S^{n-1}$ is a **good direction** for a set $T \subseteq R^n$ if each line with direction u intersects T in only finitely many points. (A line with direction u is a set $p + Ru$ with $p \in R^n$.) Otherwise u is a **bad** direction for T.

If T is definable and $\dim(T) < n$ then it follows from the good directions lemma (4.2) in Chapter 7 that almost all unit vectors $u \in S^{n-1}$ are good directions for T, in the sense that the set of bad directions is of dimension $< n - 1$.

(1.4) LEMMA. *Let $T \subseteq R^{m+n}$ be definable, $\dim(T) < m + n$, $n > 0$. Then there exists a unit vector $u \in S^{n-1}$ such that the set*

$$\{a \in R^m : u \text{ is a bad direction for } T_a\}$$

has dimension $< m$, that is, u is a good direction for almost all sets T_a.

Proof. Note that $\dim\{a \in R^m : \dim(T_a) = n\} < m$, so we may as well discard from T the subsets $\{a\} \times T_a$ for which $\dim(T_a) = n$. Hence we have reduced to the case that $\dim(T_a) < n$ for all $a \in R^m$. For each $a \in R^m$ let $B_a \subseteq S^{n-1}$ be the set of bad directions for T_a. Then $\dim(B_a) < n-1$ for all $a \in R^m$, so $\dim(B) < m+n-1$ where $B := \bigcup\{\{a\} \times B_a : a \in R^m\} \subseteq R^m \times S^{n-1}$. Let $p: R^m \times S^{n-1} \to S^{n-1}$ be the projection map onto the second factor. Then it follows from Chapter 4, Section 1 that there is $u \in S^{n-1}$ such that $\dim(p^{-1}(u) \cap B) < m$. This vector u is a good direction for almost all T_a. □

The following technical lemma is the key to the trivialization theorem. To formulate it we need a mild generalization of cells: define a **generalized open cell in R^m** to be a definable open subset of R^m that is definably homeomorphic to R^m. Examples are open cells in R^m and m-simplexes in R^m. Clearly each generalized open cell is definably contractible to each of its points.

(1.5) Lemma. *Let $S_1, \ldots, S_k \subseteq R^{m+n}$ be definable, let $A \subseteq R^m$ be definable, and let $\pi : R^{m+n} \to R^m$ be the projection map. Then there are disjoint generalized open cells $A_1, \ldots, A_h \subseteq A$ such that $\dim(A - \bigcup A_i) < m$, and such that for each $i \in \{1, \ldots, h\}$ and each definable contraction $H : A_i \times [0,1] \to A_i$ from A_i to a point $a \in A_i$ there exist a definable retraction $r : A_i \times R^n \to \{a\} \times R^n$ and a definable contraction $\tilde{H} : A_i \times R^n \times [0,1] \to A_i \times R^n$ from $A_i \times R^n$ to r, with the following properties:*

(1) *$\tilde{H}$ lifts H, that is, the following diagram commutes:*

$$\begin{array}{ccc} A_i \times R^n \times [0,1] & \xrightarrow{\tilde{H}} & A_i \times R^n \\ {\scriptstyle \pi_i \times id}\downarrow & & \downarrow{\scriptstyle \pi_i} \\ A_i \times [0,1] & \xrightarrow[H]{} & A_i \end{array}$$

where $\pi_i = \pi | A_i \times R^n$.

(2) *For each S_j the maps r and $\tilde{H}$ yield by restriction a definable retraction*

$$r_j : (A_i \times R^n) \cap S_j \to (\{a\} \times R^n) \cap S_j$$

and a definable contraction

$$\tilde{H}_j : ((A_i \times R^n) \cap S_j) \times [0,1] \to (A_i \times R^n) \cap S_j$$

from $(A_i \times R^n) \cap S_j$ to r_j.

(3) *For each $x \in A_i$ the map r restricts to a definable homeomorphism*

$$r_x : \{x\} \times R^n \to \{a\} \times R^n$$

that maps $(\{x\} \times R^n) \cap S_j$ onto $(\{a\} \times R^n) \cap S_j$ for each j.

PROOF. By induction on n. The case $n = 0$ follows by cell decomposition. In more detail: take a finite partition of A into cells that is compatible with $S_1, \dots, S_k$, and let $A_1, \dots, A_h$ be the open cells in this partition; then H and the point $a \in A_i$ determine r, and by taking $\tilde{H} = H$ the conditions (1), (2), and (3) are trivially satisfied.

Let $n > 0$ and assume inductively that the assertion holds for $n - 1$ instead of n. We then first do the case that $S_1, \dots, S_k$ are all *bounded*. Let

$$T := \bigcup_j \mathrm{bd}(S_j),$$

so that T is closed and bounded, and $\dim(T) < m + n$. (Later we modify T to a set $\tilde{T}$ that is the graph of a full multivalued function.) By the previous lemma there is a unit vector $u \in S^{n-1}$ that is a good direction for almost all sets T_a, that is, $\dim(B) < m$ where $B := \{a \in R^m : u \text{ is a bad direction for } T_a\}$.

Recall that $e_1, \dots, e_n$ denotes the standard basis of the vector space R^n over R. We view e_n as pointing in the "vertical direction". Take a linear automorphism $\lambda : R^n \to R^n$ such that $\lambda(u) = e_n$, and replace each set S_j by its image $\{(x, \lambda(y)) : (x, y) \in S_j\}$ under the automorphism $id \times \lambda$ of R^{m+n}. This is permitted since a solution to our problem for the new S_j's can be transformed back via $id \times \lambda^{-1}$ to a solution of the problem for the original S_j's. Note also that then T gets replaced by $\{(x, \lambda(y)) : (x, y) \in T\}$, so our new T satisfies

$$\dim(B) < m, \text{ where } B = \{a \in R^m : e_n \text{ is a bad direction for } T_a\}.$$

Now let $p : R^{m+n} \to R^{m+n-1}$ be the projection map on the first $m+n-1$ coordinates and $q : R^{m+n-1} \to R^m$ the projection map on the first m coordinates, so that $\pi = q \circ p$. Then e_{m+n} is a good direction for $T \cap \pi^{-1}(A - B)$. Hence by cell decomposition the set $T \cap \pi^{-1}(A - B)$ is the disjoint union of $\Gamma(f)$'s for finitely many definable continuous functions f defined on cells C that form a finite partition $\mathcal{P}$ of $(A - B) \times R^{n-1}$. Call these functions on cells $C \in \mathcal{P}$ **distinguished**. Take a triangulation (Φ, K) of $(A - B) \times R^{n-1}$ compatible with each cell $C \in \mathcal{P}$. Let F be the collection of the restrictions $f|D$ of the distinguished functions f on cells $C \in \mathcal{P}$ to sets $D \in \Phi^{-1}(K)$ with $D \subseteq C$. Then F is a multivalued function on $((A-B) \times R^{n-1}, \Phi^{-1}(K))$ in the sense of (2.5) of Chapter 8, and F is closed by (2.6) of Chapter 8, since $\Gamma(F) = T \cap \pi^{-1}(A - B)$ is closed in $\pi^{-1}(A - B) = (A - B) \times R^n$. Unfortunately, F may not be full, and therefore we modify F as follows to a full multivalued function $\tilde{F}$. First extend each $f \in F|D$ continuously to the closure $\mathrm{cl}(D) \cap (A - B)$ of D in $A - B$, and then further to a bounded continuous definable function $\tilde{f} : A - B \to R$, using (3.10) of Chapter 8. Now put

$$\tilde{T} := \left(\bigcup_{f \in F} \Gamma(\tilde{f}) \right) \cup \Gamma(f_0),$$

where $f_0 : (A - B) \times R^{n-1} \to R$ is the function that is identically 0.

Next replace (Φ, K) by a new triangulation $(\tilde{\Phi}, \tilde{K})$ of $(A - B) \times R^{n-1}$ that is compatible with all sets in $\Phi^{-1}(K)$, all sets $p(S_j \cap \Gamma(\tilde{f}))$ for $f \in F$, and all sets $p(S_j \cap \Gamma(f_0))$. In addition we arrange that the nonempty restrictions of the $\tilde{f}$'s for $f \in F$ and of f_0 to the sets in $\tilde{\Phi}^{-1}(\tilde{K})$ form a multivalued function $\tilde{F}$ on the triangulated set $((A - B) \times R^{n-1}, \tilde{\Phi}^{-1}(\tilde{K}))$.

Note that $\Gamma(\tilde{F}) = \tilde{T}$, that $\tilde{T}$ is closed in $(A-B)\times R^n$, and that $T\cap((A-B)\times R^n) \subseteq \tilde{T}$. Clearly $\tilde{F}$ is full. Let $\tilde{F}|D = \{f_{D1}, \ldots, f_{Dk(D)}\}$ with $f_{D1} < \cdots < f_{Dk(D)}$, $k(D) \geq 1$, where $D \in \tilde{\Phi}^{-1}(\tilde{K})$. The compatibility of $\tilde{\Phi}^{-1}(\tilde{K})$ with the sets indicated, together with (3.3) of Chapter 6, implies that for any $f, g \in \tilde{F}|D$ and set S_j we have:

$$(*) \qquad \begin{cases} \text{either } \Gamma(f) \subseteq S_j, \text{ or } \Gamma(f) \cap S_j = \emptyset; \\ \text{if } f < g \text{ are successive members of } \tilde{F}|D, \\ \text{then either } (f,g) \subseteq S_j, \text{ or } (f,g) \cap S_j = \emptyset; \\ \text{if } f = f_{D1} \text{ is the first member of } \tilde{F}|D, \text{ then } (-\infty, f) \cap S_j = \emptyset; \\ \text{if } g = f_{Dk(D)} \text{ is the last member of } \tilde{F}|D, \text{ then } (g, +\infty) \cap S_j = \emptyset. \end{cases}$$

Now apply the inductive hypothesis to the sets in $\tilde{\Phi}^{-1}(\tilde{K})$ and $A - B$ instead of A. This gives disjoint generalized open cells $A_1, \ldots, A_h \subseteq A - B$ with $\dim(A - \bigcup A_i) < m$ and with the following properties for any given $i \in \{1, \ldots, h\}$:

(1)$'$ Each definable contraction $H : A_i \times [0,1] \to A_i$ from A_i to a point $a \in A_i$ can be lifted to a definable contraction $H' : A_i \times R^{n-1} \times [0,1] \to A_i \times R^{n-1}$ from $A_i \times R^{n-1}$ to a definable retraction $r' : A_i \times R^{n-1} \to \{a\} \times R^{n-1}$.

(2)$'$ For each $D \in \tilde{\Phi}^{-1}(\tilde{K})$ the maps r' and H' from (1)$'$ yield by restriction a definable retraction $r'_D : (A_i \times R^{n-1}) \cap D \to (\{a\} \times R^{n-1}) \cap D$ and a definable contraction $H'_D : [(A_i \times R^{n-1}) \cap D] \times [0,1] \to (A_i \times R^{n-1}) \cap D$ from $(A_i \times R^{n-1}) \cap D$ to r'_D.

(3)$'$ For each $x \in A_i$ the map r' from (1)$'$ restricts to a definable homeomorphism $r'_x : \{x\} \times R^{n-1} \to \{a\} \times R^{n-1}$ mapping $(\{x\} \times R^{n-1}) \cap D$ onto $(\{a\} \times R^{n-1}) \cap D$, for each $D \in \tilde{\Phi}^{-1}(\tilde{K})$.

Fix an A_i and write $A_i \times R^n$ as the disjoint union of the following sets:

$$\begin{aligned} G_{Dk} &:= \{(y, f_{Dk}(y)) : y \in (A_i \times R^{n-1}) \cap D\}, \quad D \in \tilde{\Phi}^{-1}(\tilde{K}),\ 1 \leq k \leq k(D); \\ U_{Dk} &:= \{(y, s) : y \in (A_i \times R^{n-1}) \cap D,\ f_{Dk}(y) < s < f_{Dk+1}(y)\}, \\ &\qquad D \in \tilde{\Phi}^{-1}(\tilde{K}),\ 1 \leq k < k(D); \\ U_{D-} &:= \{(y, s) : y \in (A_i \times R^{n-1}) \cap D,\ s < f_{D1}(y)\}, \quad D \in \tilde{\Phi}^{-1}(\tilde{K}); \\ U_{D+} &:= \{(y, s) : y \in (A_i \times R^{n-1}) \cap D,\ s > f_{Dk(D)}(y)\}, \quad D \in \tilde{\Phi}^{-1}(\tilde{K}). \end{aligned}$$

By $(*)$ every set $(A_i \times R^n) \cap S_j$ is a union of some of these sets.

Let a definable contraction $H : A_i \times [0,1] \to A_i$ from A_i to a point $a \in A_i$ be given, and let H' and r' be as in (1)$'$, (2)$'$ and (3)$'$ above. We now show how to lift H' to a definable homotopy $\tilde{H}: A_i \times R^n \times [0,1] \to A_i \times R^n$ by defining $\tilde{H}$ on each of the sets $G_{Dk} \times [0,1]$, $U_{Dk} \times [0,1]$, $U_{D-} \times [0,1]$ and $U_{D+} \times [0,1]$:

(i) $\tilde{H}\big(y, f_{Dk}(y), t\big) := \big(H'(y,t),\ f_{Dk}(H'(y,t))\big)$, for $\big(y, f_{Dk}(y)\big) \in G_{Dk}$, $t \in [0,1]$;

(ii) write each $z \in U_{Dk}$ uniquely as $z = \big(y, (1-u)f_{Dk}(y) + uf_{Dk+1}(y)\big)$ with $0 < u < 1$ and set, for $t \in [0,1]$,
$$\tilde{H}(z,t) := \big(H'(y,t),\ (1-u)f_{Dk}\big(H'(y,t)\big) + uf_{Dk+1}\big(H'(y,t)\big)\big);$$

(iii) write each $z \in U_{D-}$ uniquely as $z = \big(y, f_{D1}(y) - u\big)$ with $u > 0$ and set $\tilde{H}(z,t) := \big(H'(y,t), f_{D1}\big(H'(y,t)\big) - u\big)$, for $t \in [0,1]$;

(iv) write each $z \in U_{D+}$ uniquely as $z = \big(y, f_{Dk(D)}(y) + u\big)$ with $u > 0$ and set $\tilde{H}(z,t) := \big(H'(y,t), f_{Dk(D)}\big(H'(y,t)\big) + u\big)$, for $t \in [0,1]$.

Clearly $\tilde{H}$ maps $G_{Dk} \times [0,1]$ into G_{Dk}, $U_{Dk} \times [0,1]$ into U_{Dk}, $U_{D-} \times [0,1]$ into U_{D-}, and $U_{D+} \times [0,1]$ into U_{D+}. A lengthy but routine argument using fullness of $\tilde{F}$ (and for instance lemma (4.2) from Chapter 6) shows that $\tilde{H}$ is continuous. Hence $\tilde{H}$ is a definable contraction from $A_i \times R^n$ to the retraction $r: A_i \times R^n \to \{a\} \times R^n$ given by $r(z) := \tilde{H}(z,1)$. One checks easily that $\tilde{H}$ and r have the properties required in (1) and (2). To check (3) one shows first that for fixed $y \in A_i \times R^{n-1}$ we have $r(y,s) = \big(r'(y), g_y(s)\big)$, where $g_y : R \to R$ is a strictly increasing bijection. Thus each map $r_x: \{x\} \times R^n \to \{a\} \times R^n$ as in (3) is bijective. To get the continuity of the inverse of r_x one first shows there is a constant $M > 0$ in R, independent of $y \in A_i \times R^{n-1}$, such that $|g_y(s) - s| \leq M$ for all s, and then one applies lemma (4.2) from Chapter 6, and the inductive assumption (3)$'$ on r'_x.

This settles the case that $S_1, \ldots, S_k$ are bounded. Now we reduce the general case to this special case. Let I be the interval $(-1,1)$ and $\mu: R \to I$ the homeomorphism given by $\mu(x) := x/\sqrt{1+x^2}$, and let $\mu_p: R^p \to I^p$ be the homeomorphism given by $\mu_p(x_1, \ldots, x_p) := \big(\mu(x_1), \ldots, \mu(x_p)\big)$. Note that μ_p is definable.

Let definable $S_1, \ldots, S_k \subseteq R^{m+n}$ and definable $A \subseteq R^m$ be given. Then the sets $S_{\mu i} := \mu_{m+n}(S_i) \subseteq R^{m+n}$ are bounded, and we add to $S_{\mu 1}, \ldots, S_{\mu k}$ one extra set, namely $S_{\mu 0} := I^{m+n}$, so $S_{\mu 0} = \mu_{m+n}(R^{m+n})$. Put $A_\mu := \mu_m(A)$. We now apply the result proved for the bounded case to the sets $S_{\mu 0}, \ldots, S_{\mu k}$, with A_μ playing the role of A. This gives us disjoint generalized open cells $A_{\mu 1}, \ldots, A_{\mu h} \subseteq A_\mu$ with $\dim\big(A_\mu - \bigcup A_{\mu i}\big) < m$ such that for any given $i \in \{1, \ldots, h\}$ we have:

(1_μ) each definable contraction $H_\mu: A_{\mu i} \times [0,1] \to A_{\mu i}$ to a point $a_\mu \in A_{\mu i}$ lifts to a definable contraction $\tilde{H}_\mu: A_{\mu i} \times R^n \times [0,1] \to A_{\mu i} \times R^n$ from $A_{\mu i} \times R^n$ to a definable retraction $r_\mu: A_{\mu i} \times R^n \to \{a_\mu\} \times R^n$;

(2_μ) for each $S_{\mu j}$ (including $S_{\mu 0}$) the maps r_μ and $\tilde{H}_\mu$ from (1_μ) yield by restriction a definable retraction $r_{\mu j}: (A_{\mu i} \times R^n) \cap S_{\mu j} \to \big(\{a_\mu\} \times R^n\big) \cap S_{\mu j}$, and a definable contraction $\tilde{H}_{\mu j}: \big((A_{\mu i} \times R^n) \cap S_{\mu j}\big) \times [0,1] \to (A_{\mu i} \times R^n) \cap S_{\mu j}$ from $(A_{\mu i} \times R^n) \cap S_{\mu j}$ to $r_{\mu j}$;

(3_μ) for each $x \in A_{\mu i}$ the map r_μ from (1_μ) restricts to a homeomorphism $r_{\mu x}$: $\{x\} \times R^n \to \{a_\mu\} \times R^n$ that maps $\big(\{x\} \times R^n\big) \cap S_{\mu j}$ onto $\big(\{a_\mu\} \times R^n\big) \cap S_{\mu j}$ for each j.

Now we use $\mu_m^{-1} : I^m \to R^m$ and $\mu_{m+n}^{-1} : I^{m+n} \to R^{m+n}$ to transform back. Put $A_i := \mu_m^{-1}(A_{\mu i})$, $i = 1,\ldots,h$. Clearly $A_1,\ldots,A_h$ are generalized open cells and $\dim(A - \bigcup A_i) < m$. Let $H : A_i \times [0,1] \to \{a\}$ be a definable contraction from A_i to a point $a \in A_i$. We shall find r and $\tilde{H}$ with properties (1), (2), and (3) of the lemma. Set $a_\mu := \mu_m(a)$ and define $H_\mu : A_{\mu i} \times [0,1] \to A_{\mu i}$ by $H_\mu(x,t) := \mu_m H(\mu_m^{-1}(x), t)$. Then H_μ is a definable contraction from $A_{\mu i}$ to a_μ. Let $\tilde{H}_\mu$ and r_μ be as in (1_μ), (2_μ), and (3_μ) above. Applying (2_μ) to $S_{\mu 0} = I^{m+n}$ we see that $\tilde{H}_\mu$ maps $A_{\mu i} \times I^n \times [0,1]$ into $A_{\mu i} \times I^n$, so we may define $\tilde{H} : A_i \times R^n \times [0,1] \to A_i \times R^n$ by

$$\tilde{H}(x,y,t) := \mu_{m+n}^{-1} \tilde{H}_\mu(\mu_m(x), \mu_n(y), t).$$

Similarly r_μ maps $A_{\mu i} \times I^n$ into $\{a_\mu\} \times I^n$, by $(2'_\mu)$ applied to $S_{\mu 0}$, so we may define $r : A_i \times R^n \to \{a\} \times R^n$ by $r(x,y) := \mu_{m+n}^{-1} r_\mu(\mu_m(x), \mu_n(y))$. One verifies easily that then $\tilde{H}$ and r satisfy the statements of the lemma. □

(1.6) In our trivialization theorem we also want to take into account distinguished definable subsets of S. Here are the relevant definitions.

Let $A \subseteq R^m$ and $S \subseteq R^n$ be definable sets, $f : S \to A$ a continuous definable map, and $S_1,\ldots,S_k$ definable subsets of S.

Then a **definable trivialization of f respecting** $S_1,\ldots,S_k$ consists of a definable trivialization (F,λ) of f and distinguished definable subsets $F_1,\ldots,F_k$ of F, such that $(f,\lambda)(S_i) = A \times F_i$ for $i = 1,\ldots,k$. Note that then each restriction $f|S_i : S_i \to A$ has definable trivialization $(F_i, \lambda|S_i)$. Given definable $A' \subseteq A$ we call $(f; S_1,\ldots,S_k)$ **definably trivial over** A' if the restriction

$$f|f^{-1}(A') : f^{-1}(A') \to A'$$

has a definable trivialization respecting $S_1 \cap f^{-1}(A'),\ldots,S_k \cap f^{-1}(A')$. Note that then $(f; S_1,\ldots,S_k)$ is also definably trivial over each definable subset of A'. Also, if $a \in A'$ we can always take a definable trivialization over A' respecting $S_1 \cap f^{-1}(A'),\ldots,S_k \cap f^{-1}(A')$ to be of the form $(f^{-1}(a), \lambda)$ with λ the identity on $f^{-1}(a) \subseteq f^{-1}(A')$, and with distinguished subsets $S_1 \cap f^{-1}(a),\ldots,S_k \cap f^{-1}(a)$ of $f^{-1}(a)$.

(1.7) TRIVIALIZATION THEOREM. *With $f : S \to A$ and $S_1,\ldots,S_k$ as above, we can partition A into definable subsets $A_1,\ldots,A_M$, such that $(f; S_1,\ldots,S_k)$ is definably trivial over each A_i.*

PROOF. By induction on $\dim(A)$. If $\dim(A) \le 0$, then A is finite, and the theorem holds trivially. Let $\dim(A) > 0$, and assume the desired result holds for lower values of $\dim(A)$. By partitioning A into finitely many cells A', and working with the restrictions $f|f^{-1}(A') : f^{-1}(A') \to A'$ we may as well assume that A is a cell; then using the homeomorphism p_A from A onto an open cell, we may further assume A is an open cell, say in R^m, so $\dim(A) = m$. Next we replace S by the reversed graph of f by applying the definable homeomorphism $y \mapsto (f(y), y)$ from S onto this reversed graph. So from now on S is a definable subset of R^{m+n}, and f is the restriction $\pi|S : S \to A$ of the projection map $\pi : R^{m+n} \to R^m$. We need one further reduction to the bounded case. Let $\mu : R \to I = (-1,1)$ be the definable homeomorphism given by $\mu(x) = x/\sqrt{1+x^2}$, and let $\mu_p : R^p \to I^p$ be the induced homeomorphism given by $\mu_p(x_1, \ldots, x_p) = (\mu(x_1), \ldots, \mu(x_p))$, for $p \in \mathbf{N}$. Put $S_\mu := \mu_{m+n}(S)$, $S_{\mu j} := \mu_{m+n}(S_j)$, $A_\mu := \mu_m(A)$, and let $f_\mu : S_\mu \to A_\mu$ be the restriction of the projection map $R^{m+n} \to R^m$. Clearly it suffices to show

$$(*) \quad \begin{cases} A_\mu \text{ can be partitioned into definable subsets } A_{\mu 1}, \ldots, A_{\mu M} \text{ such that} \\ (f_\mu; S_{\mu 1}, \ldots, S_{\mu k}) \text{ is definably trivial over each } A_{\mu i}. \end{cases}$$

We distinguish one more subset of R^{m+n}, namely $A_\mu \times \mathrm{cl}(I)^n$, where $I = (-1,1)$, and apply lemma (1.5), with $S_\mu, S_{\mu 1}, \ldots, S_{\mu k}$ and $A_\mu \times \mathrm{cl}(I)^n$ as distinguished subsets of R^{m+n}, and A_μ in the role of A: this gives disjoint generalized open cells $A_{\mu 1}, \ldots, A_{\mu h}$ in A_μ with $\dim\left(A_\mu - \bigcup A_{\mu i}\right) < m$, and further properties from lemma (1.5). By the inductive assumption we can partition $A_\mu - \bigcup A_{\mu i}$ into finitely many definable subsets over each of which $(f_\mu; S_{\mu 1}, \ldots, S_{\mu k})$ is definably trivial. So $(*)$ will follow from

$$(**) \qquad (f_\mu; S_{\mu 1}, \ldots, S_{\mu k}) \text{ is trivial over each } A_{\mu i},\ 1 \le i \le h.$$

Fix an i and take a definable contraction H from $A_{\mu i}$ to one of its points a. Then lemma (1.5) gives us a definable retraction $r : A_{\mu i} \times R^n \to \{a\} \times R^n$, which for each $x \in A_{\mu i}$ maps $\{x\} \times R^n$ homeomorphically onto $\{a\} \times R^n$, $\{x\} \times S_{\mu x}$ onto $\{a\} \times S_{\mu a}$, $\{x\} \times S_{\mu j x}$ onto $\{a\} \times S_{\mu j a}$ $(1 \le j \le k)$, and $\{x\} \times \mathrm{cl}(I)^n$ onto $\{a\} \times \mathrm{cl}(I)^n$.

Write $r(x,y) = (a, s(x,y))$, where $s : A_{\mu i} \times R^n \to R^n$ is a definable continuous map. Then the map $(x,y) \mapsto (x, s(x,y)) : A_{\mu i} \times \mathrm{cl}(I)^n \to A_{\mu i} \times \mathrm{cl}(I)^n$ is a continuous definable bijection. This bijection must be a homeomorphism: each point in $A_{\mu i}$ has a definable neighborhood N in $A_{\mu i}$ that is closed and bounded, and this bijection maps $N \times \mathrm{cl}(I)^n$ homeomorphically onto itself, by Chapter 6, (1.12). Define $\lambda_i : f_\mu^{-1}(A_{\mu i}) \to S_{\mu a}$ by $\lambda_i(x,y) := s(x,y)$. It is now easy to verify that $(S_{\mu a}, \lambda_i)$ is a definable trivialization of $(f_\mu; S_{\mu 1}, \ldots, S_{\mu k})$ over $A_{\mu i}$. This finishes the proof of $(**)$. □

(1.8) Note that theorem (1.2) is the special case $k = 0$ of theorem (1.7). Readers familiar with elementary model theory will note that the easy model-theoretic arguments (in Chapter 8, (2.14)) proving theorem (1.2) can be used to derive theorem (1.7) in a similar way.

(1.9) The following exercise gives an alternative formulation of the trivialization theorem, and explains why such results are also referred to as "generic" triviality theorems.

EXERCISE. Let $f : S \to A$ be a continuous definable map between nonempty definable sets S and A. Show that there is a definable set $E \subseteq A$ with $\dim E < \dim A$ such that f is definably trivial over each definably connected component of $A - E$.

§2. Applications

(2.1) An immediate application of the trivialization theorem is that for a given definable set $S \subseteq R^{m+n}$ the definable sets S_x fall into only finitely many different definable homeomorphism types as x ranges over R^m.

To see this, let $f: S \to R^m$ be the restriction of the projection map $R^{m+n} \to R^m$, and take a partition of R^m into definable sets $A_1, \ldots, A_M$ over each of which f is definably trivial, say with definable trivialization (λ_i, F_i) over A_i. Then S_x is definably homeomorphic to F_i for all $x \in A_i$.

A (well-known) special case of this finiteness result is the fact that for fixed natural numbers d and n the zero sets of polynomials $f(X_1, \ldots, X_n) \in \mathbb{R}[X_1, \ldots, X_n]$ in $\mathbb{R}^n$ of degree at most d fall into only finitely many semialgebraic homeomorphism types. (Instead of $\mathbb{R}$ we can take here of course any real closed field.)

We can strengthen these results by considering "embedded homeomorphism type". Two sets $X, Y \subseteq \mathbb{R}^n$ have the same **embedded homeomorphism type** if there is a homeomorphism $h: \mathbb{R}^n \xrightarrow{\sim} \mathbb{R}^n$ such that $h(X) = Y$. To have the same embedded homeomorphism type is an equivalence relation on the collection of subsets of $\mathbb{R}^n$. Similarly we say that two definable sets $X, Y \subseteq R^n$ have the same **embedded definable homeomorphism type** if there is a definable homeomorphism $h: R^n \xrightarrow{\sim} R^n$ such that $h(X) = Y$. To have the same embedded definable homeomorphism type is an equivalence relation on the collection of definable subsets of R^n. Using the "trivialization theorem with distinguished subsets" the same arguments as above show

PROPOSITION. *If $S \subseteq R^{m+n}$ is definable, then the definable sets $S_x \subseteq R^n$ fall into only finitely many embedded definable homeomorphism types, as x ranges over R^m.*

(2.2) Next we give an application to the local structure of definable sets. Given a definable set $A \subseteq R^n$ and a point $p \in R^n$ not in A we define the **cone with base A and vertex p** to be the set $[A, p] := \{ta + (1-t)p : a \in A, 0 \le t \le 1\}$, the union of the line segments $[a, p]$ with $a \in A$.

Given $p \in R^n$ and $\epsilon > 0$, we also put

$$B(p, \epsilon) := \{x \in R^n : \|x - p\| \le \epsilon\}, \text{ the closed ball centered at } p \text{ with radius } \epsilon;$$
$$S(p, \epsilon) := \{x \in R^n : \|x - p\| = \epsilon\}, \text{ the sphere centered at } p \text{ with radius } \epsilon.$$

(2.3) THEOREM. *Let $E \subseteq R^n$ be a definable set and p a non-isolated point of E. Then there is $\epsilon > 0$ such that $E \cap B(p, \epsilon)$ is definably homeomorphic to the cone with base $E \cap S(p, \epsilon)$ and vertex p. More precisely, there is a definable homeomorphism ϕ from $B(p, \epsilon)$ onto itself, such that*

(i) *$\phi(p) = p$ and ϕ is the identity on $S(p, \epsilon)$,*

(ii) $\|\phi(x) - p\| = \|x - p\|$ *for all* $x \in B(p,\epsilon)$,
(iii) $\phi(E \cap B(p,\epsilon)) = [E \cap S(p,\epsilon), p]$.

PROOF. Apply the trivialization theorem to the distance function $x \mapsto \|x - p\| : R^n \to [0,\infty)$, with E as a distinguished subset of R^n. Note that for $\epsilon > 0$ the ϵ-fiber of this map is $S(p,\epsilon)$, and that the inverse image of $(0,\epsilon]$ is $B(p,\epsilon) - \{p\}$. This gives a definable trivialization $\lambda : B(p,\epsilon) - \{p\} \to S(p,\epsilon)$ respecting E over a half-interval $(0,\epsilon]$, $\epsilon > 0$, with $\lambda|S(p,\epsilon) =$ identity.

Hence we have a definable homeomorphism

$$x \mapsto (\|x - p\|, \lambda(x)) : B(p,\epsilon) - \{p\} \to (0,\epsilon] \times S(p,\epsilon)$$

mapping $E \cap (B(p,\epsilon) - \{p\})$ onto $(0,\epsilon] \times (E \cap S(p,\epsilon))$.

Now define $\phi : B(p,\epsilon) \to B(p,\epsilon)$ by $\phi(x) = p + (\|x - p\|/\epsilon) \cdot (\lambda(x) - p)$ for $x \neq p$, and $\phi(p) = p$. One verifies easily that this ϕ has all the desired properties. □

§3. On a conjecture of Benedetti and Risler

(3.1) The generic triviality theorem leads to a new result on zero sets of "fewnomials", as recorded in the following proposition. We will make essential use of the o-minimality of the model-theoretic structure $\mathbb{R}_{\exp} := (\mathbb{R}, <, 0, 1, +, -, \cdot, \exp)$ due to Wilkie [64]; see also [22] for another proof that $\mathbb{R}_{\exp}$ is o-minimal.

(3.2) PROPOSITION. *For any given natural numbers m and n there are only finitely many homeomorphism types among the sets $Z(f) := \{x \in \mathbb{R}^n : f(x) = 0\}$, where $f(X_1, \ldots, X_n) \in \mathbb{R}[X_1, \ldots, X_n]$ has at most m monomials.*

PROOF. We are of course going to use the fact that the function

$$(x, y) \mapsto x^y := \exp(y \log(x)) : (0,\infty) \times \mathbb{R} \to \mathbb{R}$$

is definable in $\mathbb{R}_{\exp}$. A minor complication is that, on the other hand, there is no function $F : \mathbb{R}^2 \to \mathbb{R}$ definable in $\mathbb{R}_{\exp}$ such that $F(x, k) = x^k$ for all $x \in \mathbb{R}$ and $k \in \mathbb{N}$. (See exercise (3.9).) Fortunately we can instead define two functions $E, O : \mathbb{R}^2 \to \mathbb{R}$ in $\mathbb{R}_{\exp}$ such that $E(x, k) = x^k$ for all $x \in \mathbb{R}$ and $k \in 2\mathbb{N}$ and $O(x, k) = x^k$ for all $x \in \mathbb{R}$ and $k \in 2\mathbb{N} + 1$. We define E and O as follows:

$$E(x,y) := \begin{cases} |x|^y & \text{for } x \neq 0, \\ 1 & \text{for } x = y = 0, \\ 0 & \text{for } x = 0, y \neq 0; \end{cases}$$

$$O(x,y) := \begin{cases} x^y & \text{for } x > 0, \\ -(-x)^y & \text{for } x < 0, \\ 0 & \text{for } x = 0. \end{cases}$$

Let us now fix natural numbers m and n, and let $f = \sum_{i=1}^{m} a_i X^{\alpha}$ denote an arbitrary polynomial in $X = (X_1, \ldots, X_n)$ over $\mathbb{R}$ with at most m monomials: $a_i \in \mathbb{R}$ and $\alpha_i = (\alpha_{i1}, \ldots, \alpha_{in}) \in \mathbf{N}^n$ for $i = 1, \ldots, m$. Let

$$\mathrm{Even}_i(f) := \{j : 1 \leq j \leq n, \alpha_{ij} \in 2\mathrm{N}\} \text{ and } \mathrm{Odd}_i(f) := \{1, \ldots, n\} - \mathrm{Even}_i(f)$$

for $i = 1, \ldots, m$. Then we have for all $x \in \mathbb{R}^n$

$$f(x) = \sum a_i \prod_{j \in \mathbf{Even}_i(f)} E(x_j, \alpha_{ij}) \prod_{j \in \mathbf{Odd}_i(f)} O(x_j, \alpha_{ij}).$$

In this expression we may consider the $(m + mn)$-tuple

$$c(f) := (a_1, \ldots, a_m, \alpha_{11}, \ldots, \alpha_{1n}, \ldots, \alpha_{m1}, \ldots, \alpha_{mn}) \in \mathbb{R}^{m+mn}$$

as a parameter. Let $\mathbb{E}_1, \ldots, \mathbb{E}_K$, with $K = 2^{mn}$, be the different n-tuples of the form $(\mathrm{Even}_1(f), \ldots, \mathrm{Even}_m(f))$ as f varies. The considerations above show that there are sets $S_1, \ldots, S_K \subseteq \mathbb{R}^{m+mn} \times \mathbb{R}^n$, definable in $\mathbb{R}_{\mathbf{exp}}$, such that if $(\mathrm{Even}_1(f), \ldots, \mathrm{Even}_m(f)) = \mathbb{E}_k$, $1 \leq k \leq K$, then $Z(f) = (S_k)_{c(f)}$. (In other words, the zero sets of the real polynomials in $X = (X_1, \ldots, X_n)$ with at most m monomials fall into 2^{mn} different definable families, where "definable" here means "definable in $\mathbb{R}_{\mathbf{exp}}$".) The desired result now follows from the fact that among the subsets of $\mathbb{R}^n$ of the form $(S_k)_c$ $(1 \leq k \leq K, c \in \mathbb{R}^{m+mn})$ there are only finitely many homeomorphism types, by (2.1). □

(3.3) REMARK. The same argument shows that for given m, n there are only finitely many different *embedded* definable homeomorphism types among the sets $Z(f) \subseteq \mathbb{R}^n$ of proposition (3.2), where "definable" refers to definability with parameters in $\mathbb{R}_{\mathbf{exp}}$. (Here we use the proposition stated at the end of (2.1).)

(3.4) From a geometric viewpoint the "number of monomials" is not a good complexity measure for polynomials, for example, it misbehaves under linear changes of variables. Therefore Benedetti and Risler [2] introduced the more geometric notion of "additive complexity", which also makes sense for rational functions. Let us say that a rational function $f \in \mathbb{R}(X)$, with $X = (X_1, \ldots, X_n)$, has **additive complexity** $\leq k$ if there are nonzero rational functions $f_1, \ldots, f_k \in \mathbb{R}(X)$ such that

$$f_1 = a_1 X^{u_1} + b_1 X^{v_1}$$

with $a_1, b_1 \in \mathbb{R}$, $u_1 = (u_{11}, \ldots, u_{1n}) \in \mathbf{Z}^n$, $v_1 = (v_{11}, \ldots, v_{1n}) \in \mathbf{Z}^n$, and for $1 < j \leq k$

$$f_j = a_j X^{u_j} \prod_{i=1}^{j-1} f_i^{r_{ji}} + b_j X^{v_j} \prod_{i=1}^{j-1} f_i^{s_{ji}},$$

with $a_j, b_j \in \mathbb{R}$, $u_j = (u_{j1}, \ldots, u_{jn}) \in \mathbf{Z}^n$, $v_j = (v_{j1}, \ldots, v_{jn}) \in \mathbf{Z}^n$, $r_{ji} \in \mathbf{Z}$, $s_{ji} \in \mathbf{Z}$ for $1 \leq i < j$, and

$$f = cX^w \prod_{j=1}^{k} f_j^{t_j}$$

with $c \in \mathbf{R}$, $w = (w_1, \ldots, w_n) \in \mathbf{Z}^n$ and $t_1, \ldots, t_k \in \mathbf{Z}$. In other words, starting with the variables $X_1, \ldots, X_n$ and real constants and allowing addition, multiplication and division one can obtain f using at most k additions (and an unlimited number of multiplications and divisions.)

Let us say that a semialgebraic set $A \subseteq \mathbf{R}^n$ has **additive complexity at most** $(n, p, k) \in \mathbf{N}^3$ if

$$(*) \quad \begin{cases} A = \bigcup_{i \in I} A_i, \text{ where } I \text{ is a finite index set and for each } i \in I \\ A_i = \{x \in \mathbf{R}^n : f_1^i(x) = \cdots = f_{q_i}^i(x) = 0, \ g_1^i(x) > 0, \ldots, g_{r_i}^i(x) > 0\}, \\ \text{with polynomials } f_\kappa^i, g_\lambda^i \in \mathbf{R}[X], \text{ all of additive complexity at most } k, \\ \text{and } \sum_{i \in I} (q_i + r_i) \leq p. \end{cases}$$

Then we have the following finiteness result conjectured by Benedetti and Risler, see pp. 214–215 of [2].

(3.5) THEOREM. *Given any triple $(n, p, k) \in \mathbf{N}^3$ there are only finitely many different embedded homeomorphism types of semialgebraic subsets of $\mathbf{R}^n$ of additive complexity at most (n, p, k).*

Note that this generalizes proposition (3.2) above on zero sets of fewnomials. As with that proposition, the theorem we just stated follows from the fact that, given (n, p, k), there are sets $S_1, \ldots, S_M \subseteq \mathbf{R}^{N+n}$ (for suitable M, N depending on (n, p, k)), each definable in $\mathbf{R}_{\exp}$, such that each semialgebraic set $A \subseteq \mathbf{R}^n$ of additive complexity at most (n, p, k) is of the form $A = (S_m)_c$ for some $m \in \{1, \ldots, M\}$ and some $c \in \mathbf{R}^N$. To prove this fact we may assume that not only (n, p, k) is given, but also the finite index set I and the natural numbers q_i $(i \in I)$ and r_i $(i \in I)$ appearing in the representation $(*)$ of the semialgebraic sets A we are considering. The desired result will then follow easily from the next two lemmas by the same kind of argument as used in the proof of (3.2).

(3.6) LEMMA. *A rational function $f \in \mathbf{R}(X)$ has additive complexity $\leq k$ if and only if there is a sequence of nonzero polynomials $g_1, \ldots, g_k \in \mathbf{R}[X]$ such that*

(i)

$$g_1 = a_1 X^{\alpha_1} + b_1 X^{\beta_1},$$

with $a_1, b_1 \in \mathbf{R}$, $\alpha_1 = (\alpha_{11}, \ldots, \alpha_{1n}) \in \mathbf{N}^n$, $\beta_1 = (\beta_{11}, \ldots, \beta_{1n}) \in \mathbf{N}^n$,

(ii) *for $1 < j \leq k$,*

$$g_j = a_j X^{\alpha_j} \prod_{i=1}^{j-1} g_i^{d_{ji}} + b_j X^{\beta_j} \prod_{i=1}^{j-1} g_i^{e_{ji}},$$

with $a_j, b_j \in \mathbf{R}$, $\alpha_j = (\alpha_{j1}, \ldots, \alpha_{jn}) \in \mathbf{N}^n$, $\beta_j = (\beta_{j1}, \ldots, \beta_{jn}) \in \mathbf{N}^n$, $d_{ji} \in \mathbf{N}$, $e_{ji} \in \mathbf{N}$ for $1 \leq i < j$, and

(iii)

$$f \cdot X^{\gamma} \prod_{j=1}^{k} g_j^{r_j} = cX^{\delta} \prod_{j=1}^{k} g_j^{s_j}$$

with $c \in \mathbb{R}$, $\gamma = (\gamma_1, \ldots, \gamma_n) \in \mathbb{N}^n$, $\delta = (\delta_1, \ldots, \delta_n) \in \mathbb{N}^n$, *and* $r_1, \ldots, r_k$, $s_1, \ldots, s_k \in \mathbb{N}$.

PROOF. If such a sequence $g_1, \ldots, g_k$ exists, then f is clearly of additive complexity $\leq k$. Suppose f is of additive complexity $\leq k$, and let $f_1, \ldots, f_k$ be a sequence of nonzero rational functions as in (3.4). By writing the integer exponents in the expressions in (3.4) as differences of natural numbers and clearing denominators one obtains a sequence $g_1, \ldots, g_k$ of nonzero polynomials in $\mathbb{R}[X]$ satisfying (i) and (ii) such that

$$f_j = g_j \Big/ X^{\epsilon_j} \prod_{i=1}^{j-1} g_i^{p_{ji}}$$

for $1 \leq j \leq k$, with $\epsilon_j = (\epsilon_{j1}, \ldots, \epsilon_{jn}) \in \mathbb{N}^n$ and $p_{ji} \in \mathbb{N}$ for $1 \leq i < j$. Then f has the form described in (iii). □

(3.7) LEMMA. *Suppose $f \in \mathbb{R}[X]$ has additive complexity $\leq k$. Let $Y = (Y_1, \ldots, Y_k)$ be a tuple of new variables. Then there are polynomials $h_1(X,Y), \ldots, h_k(X,Y) \in \mathbb{R}[X,Y]$ such that each h_j has at most three monomials, and there are two monomials $\mu_1(X,Y)$ and $\mu_2(X,Y)$ and a real constant c such that:*

(i) *for each $x \in \mathbb{R}^n$ there is a unique $y \in \mathbb{R}^k$ with $h_1(x,y) = \cdots = h_k(x,y) = 0$;*
(ii) *for each $x \in \mathbb{R}^n$, if $y \in \mathbb{R}^k$ and $h_1(x,y) = \cdots = h_k(x,y) = 0$, then $\mu_1(x,y)f(x) = c\mu_2(x,y)$;*
(iii) *the set*

$$\{x \in \mathbb{R}^n : \text{ if } y \in \mathbb{R}^k \text{ and } h_1(x,y) = \cdots = h_k(x,y) = 0, \text{ then } \mu_1(x,y) \neq 0\}$$

is open and dense in $\mathbb{R}^n$.

PROOF. Take polynomials $g_1, \ldots, g_k \in \mathbb{R}[X]$ as in the previous lemma. Then we put

$$h_1 := Y_1 - (a_1 X^{\alpha_1} + b_1 X^{\beta_1}),$$

$$\text{and for } 1 < j \leq k, \quad h_j := Y_j - \left(a_j X^{\alpha_j} \prod_{i=1}^{j-1} Y_i^{d_{ji}} + b_j X^{\beta_j} \prod_{i=1}^{j-1} Y_i^{e_{ji}} \right).$$

Also, we put $\mu_1(X,Y) := X^{\gamma} \prod_{j=1}^{k} Y_j^{r_j}$ and $\mu_2(X,Y) := X^{\delta} \prod_{j=1}^{k} Y_j^{s_j}$ with $\gamma, \delta, r_1, \ldots, r_k$ and $s_1, \ldots, s_k$ as in the previous lemma. Clearly

$$h_1(x,y) = \cdots = h_k(x,y) = 0 \Leftrightarrow y_j = g_j(x) \text{ for } j = 1, \ldots, k,$$

from which (i) and (ii) follow immediately. Since $g_1, \ldots, g_k$ are nonzero polynomials we also get property (iii). □

(3.8) To see how the last lemma is to be applied, note that conditions (i), (ii) and (iii) definably determine the function $x \mapsto f(x): \mathbb{R}^n \to \mathbb{R}$ in terms of the polynomials $h_1, \ldots, h_k$, the monomials μ_1 and μ_2, and the constant c: it is the unique continuous extension to $\mathbb{R}^n$ of the function $x \mapsto c\mu_2(x,y)/\mu_1(x,y)$ defined on the open dense set in (iii), where $y \in \mathbb{R}^k$ is given by $h_1(x,y) = \cdots = h_k(x,y) = 0$.

(3.9) EXERCISE. Show that there is no function $F: \mathbb{R}^2 \to \mathbb{R}$, definable with parameters in $\mathbb{R}_{\exp}$, such that $F(x,k) = x^k$ for all $x \in \mathbb{R}$ and $k \in \mathbb{N}$.

Hint: use that $\mathbb{R}_{\exp}$ is o-minimal.

Notes and comments

The semialgebraic trivialization theorem over $\mathbb{R}$ is due to Hardt [29], who notes that his arguments go through for the category of bounded subanalytic sets. For similar results about the category of Nash manifolds, see Coste and Shiota [10]. Here we follow closely the treatment of the semialgebraic case over arbitrary real closed fields in Delfs and Knebusch [11]. There is a difficulty similar to the one in the previous chapter. This explains why the key lemma (1.5) was obtained in a somewhat roundabout way (first doing the bounded case, and then reducing to that case) compared to the route available for the corresponding semialgebraic result (6.3) in [11].

Subsequently (in 1993) I found the easy theorem (2.2) of Chapter 6, which leads to the short model-theoretic proof in Chapter 8, (3.13) and (2.14), that triangulation implies trivialization. But it seemed reasonable to keep also the longer but quite explicit "geometric" construction of the trivialization. The local conical structure of definable sets (theorem (2.3)) is a familiar consequence of trivialization, as in the semialgebraic case treated in [4].

As soon as Wilkie's theorem was available (in 1991) it was clear that one could use the trivialization theorem with distinguished subsets to answer the question of Risler and Benedetti as done in Section 3.

The following came too late to my attention to include in the references at the end of the book. It is closely related to the material of this chapter.
M. Coste, *Topological types of fewnomials*, preprint, December 1996.

CHAPTER 10

DEFINABLE SPACES AND QUOTIENTS

Introduction

Up till now our definable sets were always given as subsets of an ambient space R^m, a very convenient restriction that has served us well. In this final chapter we want to break out of this restricted setting, and consider also, say, projective space, and its "definable" subspaces, more generally, spaces that are not given as subsets of R^m, but locally look like definable subsets of R^m. To stay within the context of "spaces of finite type" we require a covering of the spaces of interest by only finitely many "affine" definable patches. This idea is carried out in detail in Section 1. The main result of this section, theorem (1.8), generalizes a theorem of Robson for semialgebraic spaces, to the effect that a "definable space" (obtained by gluing finitely many affine definable sets) is isomorphic to an affine definable set if and only if the space is **regular**, a separation condition that is easily verified in many situations of interest.

In Section 2 we consider a related construction, namely that of taking the quotient space X/E of a definable set X by a definable equivalence relation E on X. (Definably gluing finitely many definable sets can be viewed as a special case of this construction, but is better treated separately, as we do in Section 1.) We ask when X/E can be realized as an ordinary definable set, and are particularly interested in the case that E is "definably proper" over X. The main positive result in this direction extends a semialgebraic theorem of Brumfiel to our setting. The proof follows Brumfiel's, with some modifications. Triangulation is the key tool.

The two sections are independent. The point of this chapter is that the entire theory of the previous chapters is not restricted to just definable sets in R^m: we are free to use certain constructions on these definable sets that carry us *outside* R^m.

We fix an o-minimal expansion $\mathcal{R} = (R, <, \dots)$ of an ordered real closed field. "Definable set" will always mean "definable subset of R^m for some m", unless specified otherwise.

§1. Definable spaces

(1.1) We start by recalling the construction of a topological space by gluing: Let a covering $S = \bigcup_i U_i$ of a set S by subsets U_i $(i \in I)$ be given, and for each index $i \in I$ a set-theoretic bijection $g_i : U_i \to U_i^!$, with $U_i^!$ a topological space, such that for all i, j the set $g_i(U_i \cap U_j)$ is open in $U_i^!$ and the "transition" map

$$g_{ij} : g_i(U_i \cap U_j) \to g_j(U_j \cap U_i) \text{ with } g_{ij}(x) = g_j\big(g_i^{-1}(x)\big)$$

is continuous (hence a homeomorphism, since g_{ij} is a bijection with inverse g_{ji}). Then we equip S with the unique topology in which each U_i is open, and each $g_i : U_i \to U_i^!$ is a homeomorphism. (In this topology a subset X of S is open iff $g_i(X \cap U_i)$ is open for all $i \in I$.)

(1.2) Suppose in addition that the index set I is finite, that each $U_i^! \subseteq R^{m(i)}$ is a definable set (with the induced topology from $R^{m(i)}$), and that for each pair i, j the set $g_i(U_i \cap U_j) \subseteq U_i^!$ is definable and $g_{ij} : g_i(U_i \cap U_j) \to g_j(U_j \cap U_i)$ is definable. (Such a family $\big(g_i : U_i \to U_i^!\big)_{i \in I}$ is called a **definable atlas on** S, and the g_i's **charts** of the atlas.)

Then we extend the notion of "definable set" to subsets of S: Let $X \subseteq S$ and $f : X \to R$; call X **definable** if $g_i(X \cap U_i) \subseteq U_i^!$ is definable for each i, and call f **definable** if X is definable and $f_i : g_i(X \cap U_i) \to R$ given by $f_i(x) = f\big(g_i^{-1}(x)\big)$ is definable for each i. (We need this mainly for open X and continuous f.) Note that the definable subsets of S form a boolean algebra of subsets of S, and that for given definable $X \subseteq S$ the definable functions $f : X \to R$ form an R-algebra under pointwise addition and multiplication of functions. (The constant functions $X \to R$ are of course definable.) If $X = X_1 \cup \cdots \cup X_n$, where all X_k are definable, then a function $f : X \to R$ is definable iff each restriction $f|X_k : X_k \to R$ is definable.

Note that U_i is definable, and for $g_i = \big(g_{i1}, \ldots, g_{im(i)}\big) : U_i \to R^{m(i)}$, each g_{ik} is definable. Let $\mathrm{DO}(S)$ be the collection of definable open subsets of S, and for each $U \in \mathrm{DO}(S)$, let $\mathrm{DC}(U)$ be the R-algebra of definable continuous functions $f : U \to R$. Then we call the set S equipped with the sheaf $\big(\mathrm{DC}(U)\big)_{U \in \mathrm{DO}(S)}$ a **definable space**. This is of course only a sheaf in a *finite* sense: if $U = U_1 \cup \cdots \cup U_n$ where $U_k \in \mathrm{DO}(S)$ for all k, then a function $f : U \to R$ belongs to $\mathrm{DC}(U)$ iff $f|U_k : U_k \to R$ belongs to $\mathrm{DC}(U_k)$ for all k. The topology on S is not lost in viewing S as a definable space, since $\mathrm{DO}(S)$ is a basis for the topology. However, the particular definable atlas $\big(g_i : U_i \to U_i^!\big)_{i \in I}$ that makes S into a definable space is not recoverable from the sheaf, and different definable atlases on S can make S into the same definable space. Two definable atlases $\big(g_i : U_i \to U_i^!\big)_{i \in I}$ and $\big(h_j : V_j \to V_j^!\big)_{j \in J}$ (on the abstract set S) are called **equivalent**, if for all $i \in I$ and $j \in J$ the sets $g_i(U_i \cap V_j)$ and $g_j(V_j \cap U_i)$ are open definable subsets of $U_i^!$ and $V_j^!$ respectively, and the transition maps $x \mapsto h_j\big(g_i^{-1}(x)\big) : g_i(U_i \cap V_j) \to h_j(V_j \cap U_i)$ and their inverses $y \mapsto g_i\big(h_j^{-1}(y)\big)$ are definable and continuous. One checks easily that then (g_i) and (h_j) give rise to the same topology on S, to the same notion of definable subset of S, and to the same notion of definable function on a definable subset of S.

CONCLUSION. Two definable atlases (g_i) and (h_j) on S are equivalent iff they give rise to the same sheaf on S, that is, they make S into the same definable space. The notions of definable subset of S and definable function on a definable subset of S are now seen to depend only on S viewed as a definable space, not on the particular definable atlas used to make S into a definable space.

(1.3) Let T be a second definable space with sheaf $\big(\mathrm{DC}(V)\big)_{V\in\mathrm{DO}(T)}$. Then a **morphism** $F : S \to T$ is a map from S into T, such that if $V \in \mathrm{DO}(T)$ and $h \in \mathrm{DC}(V)$, then $F^{-1}(V) \in \mathrm{DO}(S)$ and $h \circ \big(F|F^{-1}(V)\big) \in \mathrm{DC}(F^{-1}(V))$. It follows in particular that F is continuous. If the definable space T is determined by the definable atlas $\big(h_j : V_j \to V_j^!\big)_{j\in J}$, then a map $F : S \to T$ is a morphism iff for all $i \in I$ and $j \in J$ the set $g_i\big(U_i \cap F^{-1}(V_j)\big)$ is definable and open in $U_i^!$, and the map $g_i\big(U_i \cap F^{-1}(V_j)\big) \to V_j^!$ given by $x \mapsto h_j\big(F\big(g_i^{-1}(x)\big)\big)$ is continuous and definable. Let $F : S \to T$ be a morphism. Then, given any definable set $Y \subseteq T$ and definable function $f : Y \to R$, the set $F^{-1}(Y) \subseteq S$ is definable and the function $f \circ \big(F|F^{-1}(Y)\big) : F^{-1}(Y) \to R$ is definable. Also, if $X \subseteq S$ is definable, then $F(X) \subseteq T$ is definable.

The identity map $S \to S$ is a morphism, and each constant map $S \to T$ is a morphism. The definable spaces with their morphisms form a category under the usual composition of maps.

(1.4) EXAMPLES.

(1) Each definable set $S \subseteq R^m$ is made into a definable space by taking the identity map $S \to S$ as the only chart of a definable atlas. Then the topology of the definable space S equals the topology induced by R^m, and set $X \subseteq S$ is definable in the definable space S iff it is definable as a subset of R^m, and in that case a function $f : X \to R$ is definable in the sense of the definable space S iff $\Gamma(f)$ is a definable subset of R^{m+1}. If $T \subseteq R^n$ is a second definable set, then a morphism $S \to T$ is the same thing as a continuous definable map $S \to T$. An **affine** definable space is by definition a definable space isomorphic to a definable set (in the category of definable spaces with their morphisms).

(2) (Projective spaces) Let $\mathbb{P}^n(R)$ be n-dimensional projective space over the field R; its points are the equivalence classes $(x_0 : \cdots : x_n)$ of nonzero vectors $(x_0, \ldots, x_n)$ in R^{n+1}, where two such vectors are equivalent iff one is a scalar multiple of the other. Clearly $\mathbb{P}^n(R)$ is covered by the $n+1$ subsets U_i $(0 \le i \le n)$, where U_i is the set of points $(x_0 : \cdots : x_i : \cdots : x_n)$ with $x_i \ne 0$ (or equivalently, with $x_i = 1$). Let $g_i : U_i \to R^n$ be the bijection $(x_0 : \cdots : 1 : \cdots : x_n) \mapsto (x_0, \ldots, x_{i-1}, x_{i+1}, \ldots, x_n)$. It is easy to see that $(g_i)_{0\le i\le n}$ is a definable atlas, and we make $\mathbb{P}^n(R)$ into a definable space using this atlas. In fact, $\mathbb{P}^n(R)$ is an affine definable space. To see this, define $v : \mathbb{P}^n(R) \to R^{(n+1)^2}$ by

$$v(x_0 : x_1 : \cdots : x_n) = \big(x_i x_j / (x_0^2 + \cdots + x_n^2)\big)_{0\le i,j\le n}.$$

Then v is easily seen to be an isomorphism from the definable space $\mathbb{P}^n(R)$ onto a definable subset of $R^{(n+1)^2}$.

(3) (Subspaces) Let S be a definable space given by a definable atlas $(g_i : U_i \to U_i^!)_{i\in I}$, and let $X \subseteq S$ be definable. Then $(g_i|U_i \cap X : U_i \cap X \to g_i(U_i \cap X))_{i\in I}$ is a definable atlas on the set X, and we make X into a definable space using this atlas. We call X in the role of definable space a definable subspace of S. The topology of X as a definable space is the one induced by S. The definable subsets of the definable space X are exactly the definable subsets of S that are contained in X, and given a definable set $Y \subseteq X$, a function $f: Y \to R$ is definable in the sense of the definable space X iff f is definable in the sense of S. (Thus another (equivalent) definable atlas on S makes X into the same definable space.) The inclusion $X \to S$ is a morphism. If $F: S \to T$ is a morphism between definable spaces S and T, and $F(S) \subseteq Y$, where Y is a definable subset of T, then F is also a morphism when considered as a map from S into the definable subspace Y of T.

(4) (Products) Let $S_1, \ldots, S_K$ be definable spaces, and take for each S_k a definable atlas $(g_{ki} : U_{ki} \to U'_{ki})_{i\in I_k}$, $U'_{ki} \subseteq R^{m(ki)}$, $1 \le k \le K$.

Put $I := I_1 \times \cdots \times I_K$, and for each $i = (i_1, \ldots, i_K) \in I$, let

$$\begin{aligned} U_i &:= U_{1i_1} \times \cdots \times U_{Ki_K} \subseteq S_1 \times \cdots \times S_K, \\ U_i^! &:= U'_{1i_1} \times \cdots \times U'_{Ki_K} \subseteq R^{m(1i_1)+\cdots+m(Ki_K)}, \\ g_i &:= g_{1i_1} \times \cdots \times g_{Ki_K} : U_i \to U_i^!. \end{aligned}$$

Then $(g_i : U_i \to U_i^!)_{i\in I}$ is a definable atlas on $S_1 \times \cdots \times S_K$ and we make $S_1 \times \cdots \times S_K$ into a definable space using this atlas. This gives the product topology on $S_1 \times \cdots \times S_K$. Moreover, given a definable space X and morphisms $\mu_k : X \to S_k$ for $k = 1, \ldots, K$, the map $\mu = (\mu_1, \ldots, \mu_K) : X \to S_1 \times \cdots \times S_K$ is a morphism. It follows easily that the definable space structure on the product set $S_1 \times \cdots \times S_K$ does not depend on our choice of definable atlases for the definable spaces $S_1, \ldots, S_K$.

(5) (Definable maps) Let S and T be definable spaces. Then a map $F: S \to T$ is said to be **definable** if its graph $\Gamma(F)$ is a definable subset of the product space $S \times T$, as specified in (4). One checks easily that for $T = R$ this agrees with the function $F: S \to R$ being definable in the sense of (1.2). Also $F: S \to T$ is a morphism iff F is definable and continuous.

(1.5) DEFINITION. Recall that a topological space S is said to be **regular** if for each $a \in S$ and open $U \subseteq S$ with $a \in U$ there is an open $V \subseteq S$ with $a \in V$ and $\text{cl}(V) \subseteq U$. We leave it as an exercise to show that a definable space S is regular if and only if for each $a \in S$ and definable open $U \subseteq S$ with $a \in U$ there is a definable open $V \subseteq S$ with $a \in V$ and $\text{cl}(V) \subseteq U$, and also if and only if for each $a \in S$ and definable closed $X \subseteq S$ with $a \notin X$ there are disjoint definable open neighborhoods of a and X in S.

Note that each affine definable space is regular. The theorem below states the converse. A regular topological space is clearly Hausdorff, but the following is an example due to Robson of a definable Hausdorff space S that is not regular, and hence not affine.

EXAMPLE. In the o-minimal structure $(\mathbb{R}, <, 0, 1, +, -, \cdot)$, let $U_1 = U_1' \subseteq \mathbb{R}^2$ be the union of the open unit disc and the point $p := (0,1)$. Let $U_2 = U_2' \subseteq \mathbb{R}^2$ be the open square $\{(x,y) : 0 < x < 1 \text{ and } 0 < y < 1\}$. Let $S := U_1 \cup U_2$ as a point set and consider S as a definable space via the identity mappings $g_i : U_i \longrightarrow U_i'$ for $i = 1, 2$. With these definitions, S is a definable Hausdorff space. Now consider the arc $\gamma := \{(x,y) : x^2 + y^2 = 1, x > 0, y > 0\} \subseteq U_2$: γ is closed in S since it is closed in U_2 and disjoint from U_1, while any open subset of U_2 containing γ meets U_1 in an open set whose closure contains p. Thus p and γ cannot be separated.

(1.6) DEFINITION. Let S be a definable Hausdorff space and (a,b) an interval. A definable map $\gamma : (a,b) \longrightarrow S$ is called a **definable curve** in S. For $x \in S$, we write $\gamma \to x$ to mean $\lim_{t \to b} \gamma(t) = x$. We call γ **completable** if there is a (necessarily unique) point $x \in S$ such that $\gamma \to x$. If $\gamma \to x \in X$ with definable $X \subseteq S$ such that $\gamma(a,b) \subseteq X$, we call γ **completable in** X. With these definitions one easily verifies the following facts using (4.1) and (4.2) of Chapter 6:

(1.7) LEMMA. *Let S and T be definable Hausdorff spaces, $f : S \to T$ a definable map, and $x \in S$. Then f is continuous at the point x if and only if for each definable curve γ in S with $\gamma \to x$ we have $f(\gamma) \to f(x)$, where $f(\gamma) := f \circ \gamma$.*

(1.8) THEOREM. *Every regular definable space is affine.*

PROOF. Let S be a regular definable space given by the definable atlas $(h_i : U_i \longrightarrow V_i)_{i \in I}$ with definable $V_i \subseteq R^{n_i}$. We clearly can assume each V_i is bounded. Let $l = |I|$. The case $l = 1$ is trivial. If we assume that the theorem holds for $l = 2$, then a straightforward argument proves it for all l.

We now turn to the proof of the case $l = 2$; say $I = \{1, 2\}$. If $X \subseteq R^n$, we denote by ∂X the set $\text{cl}(X) - X$. Writing $V_{12} := h_1(U_1 \cap U_2)$ and $V_{21} := h_2(U_2 \cap U_1)$, the following definable sets need to be considered:

$$B_1 := V_1 \cap \partial V_{12} = h_1(\partial U_2),$$
$$B_2 := V_2 \cap \partial V_{21} = h_2(\partial U_1),$$
$$B_1' := \{x \in R^{n_1} : \exists y \in B_2 \forall \epsilon_1, \epsilon_2 > 0\ \exists z \in U_1 \cap U_2 [d(x, h_1(z)) < \epsilon_1,\ d(y, h_2(z)) < \epsilon_2]\},$$
$$B_2' := \{y \in R^{n_2} : \exists x \in B_1 \forall \epsilon_1, \epsilon_2 > 0\ \exists z \in U_1 \cap U_2 [d(x, h_1(z)) < \epsilon_1,\ d(y, h_2(z)) < \epsilon_2]\}.$$

The following fact about these sets is crucial to the construction:

CLAIM 1. *$d(x, B_i') > 0$ for every x in V_i, $i = 1, 2$.*

PROOF. We prove the claim for $i = 1$, the case $i = 2$ being symmetrical. Assume for a contradiction that $d(x, B_1') = 0$ for some $x \in V_1$. Note that $h_1^{-1}(x) \in U_1$ and $U_1 \cap h_2^{-1}(B_2) = \emptyset$, so $U_1 \cap \operatorname{cl} h_2^{-1}(B_2) = \emptyset$. Since S is regular there are disjoint open neighborhoods D of $h_1^{-1}(x)$ in U_1 and E of $h_2^{-1}(B_2)$ in U_2. We will derive a contradiction by finding an element in $D \cap E$. Since $h_1(D)$ is open in V_1, we can take $\epsilon > 0$ such that $B(x, \epsilon) \cap V_1 \subseteq h_1(D)$. Because $d(x, B_1') = 0$ there is $x' \in B_1'$ with $d(x, x') < \epsilon$. By definition of B_1' there is $y \in B_2$ for which there are points $z \in U_1 \cap U_2$ with $h_1(z)$ arbitrarily close to x' and $h_2(z)$ arbitrarily close to y. Since $h_2(E)$ is an open neighborhood of y in V_2 it follows in particular that there is $z \in U_1 \cap U_2$ with $d(x, h_1(z)) < \epsilon$ and $h_2(z) \in h_2(E)$, hence $z \in D \cap E$. This proves the claim. □

Let $d_1(z) := d(h_1(z), B_1')$ and $d_2(z) := d(h_2(z), B_2')$ for $z \in S$. We now define the map $h: S \to R^{1+n_1+1+n_2}$ as follows:

$$h(z) := \begin{cases} \big(d_1(z), d_1(z)h_1(z), \ 0 \ , 0, \ldots, 0\big) & \text{for } z \in U_1 - U_2, \\ \big(d_1(z), d_1(z)h_1(z), \ d_2(z), d_2(z)h_2(z)\big) & \text{for } z \in U_1 \cap U_2, \\ \big(\ 0\ , 0, \ldots, 0, \ d_2(z), d_2(z)h_2(z)\big) & \text{for } z \in U_2 - U_1. \end{cases}$$

By claim 1 we clearly have for each $z \in S$

$$h(z) \in U_1 - U_2 \Leftrightarrow \text{the last } 1 + n_2 \text{ coordinates of } h(z) \text{ are equal to } 0,$$
$$h(z) \in U_2 - U_1 \Leftrightarrow \text{the first } 1 + n_1 \text{ coordinates of } h(z) \text{ are equal to } 0.$$

From this it follows easily, again using claim 1, that h is injective.

CLAIM 2. *The map h is continuous.*

To see this, let γ be a definable curve in S with $\gamma \to z \in S$. We have to show that $h(\gamma) \to h(z)$. By restricting the domain of γ suitably we may assume that γ lies either completely in $U_1 - U_2$, or completely in $U_1 \cap U_2$, or completely in $U_2 - U_1$.

CASE 1. *γ lies completely in $U_1 - U_2$.* Then also $z \in U_1 - U_2$, since $z \in U_2$ would imply that γ lies at least partly in U_2. The definition of h now gives that $h(\gamma) \to h(z)$.

CASE 2. *γ lies completely in $U_1 \cap U_2$.* If also $z \in U_1 \cap U_2$, the definition of h gives $h(\gamma) \to h(z)$. So let $z \notin U_1 \cap U_2$, say $z \notin U_1$. Then $z \in \partial U_1$, so $y := h_2(z) \in B_2$. Take $x \in R^n$, such that $h_1(\gamma) \to x$. Since $h_2(\gamma) \to y \in B_2$ it follows that $x \in B_1'$, hence $h(\gamma) \to h(z)$ as the formula above shows.

The case that γ lies completely in $U_2 - U_1$ is symmetric to case 1. This finishes the proof of claim 2. □

CLAIM 3. *h maps S homeomorphically onto $h(S)$.*

To prove this, let K be an arbitrary definable closed subset K of S. In view of the injectivity of h and claim 2 it suffices to show that $h(K)$ is then closed in $h(S)$. Let $z \in S$ with $h(z) \in \mathrm{cl}(h(K)) \cap h(S)$; it is enough that we derive from this that $z \in K$. Since $h(z) \in \mathrm{cl}(h(K))$ there is a definable curve γ in K such that $h(\gamma) \to h(z)$. We may assume that γ lies either completely in $U_1 - U_2$, or completely in $U_1 \cap U_2$, or completely in $U_2 - U_1$. We may also assume the domain of γ is an interval $(0, \epsilon)$.

CASE 1. *The curve γ lies completely in $U_1 - U_2$.* Since

$$h(\gamma(t)) = (d_1(\gamma(t)), d_1(\gamma(t))h_1(\gamma(t)), 0, 0, \ldots, 0) \to h(z) \text{ as } t \to \epsilon,$$

the last $1 + n_2$ coordinates of $h(z)$ are 0, so that $z \in U_1 - U_2$, and

$$h(z) = (d_1(z), d_1(z)h_1(z), 0, 0, \ldots, 0).$$

Using claim 1 this gives $h_1(\gamma) \to h_1(z)$, hence $\gamma \to z$, so $z \in K$, as K is closed in S.

CASE 2. *The curve γ lies completely in $U_1 \cap U_2$.* Then $h(\gamma(t)) \to h(z)$ as $t \to \epsilon$, where

$$h(\gamma(t)) = (d_1(\gamma(t)), d_1(\gamma(t))h_1(\gamma(t)), d_2(\gamma(t)), d_2(\gamma(t))h_2(\gamma(t))).$$

If $z \in U_1$, then $h(z) = (d_1(z), d_1(z)h_1(z), \ldots)$, so that by claim 1 we obtain $h_1(\gamma) \to h_1(z)$, and thus $\gamma \to z$ (since h_1 is a homeomorphism onto V_1), hence $z \in K$. If $z \in U_2$ the argument uses instead the last n_2 coordinates of $h(z)$.

The case that γ lies completely in $U_2 - U_1$ is symmetric to case 1. This finishes the proof of claim 3, and thereby the proof of the theorem. □

§2. Definable quotient spaces

(2.1) Given a definable map $f: X \to Y$ between definable sets X and Y, we denote by E_f the kernel of f, that is, $E_f = \{(x, y) \in X \times X : f(x) = f(y)\}$, a definable equivalence relation on the set X. Note that if f is continuous, then E_f is closed in $X \times X$.

(2.2) Let $E \subseteq X \times X$ be a definable equivalence relation on a definable set $X \subseteq R^m$.

DEFINITION. A **definable quotient of X by E** is a pair (p, Y) consisting of a definable set $Y \subseteq R^n$ and a definable continuous surjective map $p: X \to Y$ such that:

(i) $E = E_p$, that is $(x_1, x_2) \in E \Leftrightarrow p(x_1) = p(x_2)$, for all $x_1, x_2 \in X$;
(ii) p is definably identifying, that is, for all definable $K \subseteq Y$, if $p^{-1}(K)$ is closed in X, then K is closed in Y. (See Chapter 6, (4.4).)

(2.3) REMARKS.

(1) If clause (ii) above is replaced by the stronger clause

"p is definably proper",

then we call (p, Y) a **definably proper quotient of X by E**. (In the presence of surjectivity "definably proper" implies "definably identifying", see Chapter 6, (4.6).)

(2) If (p, Y) and (p', Y') are both definable quotients of X by E, then by (i) we clearly have a unique definable bijection $h: Y \to Y'$ such that the diagram

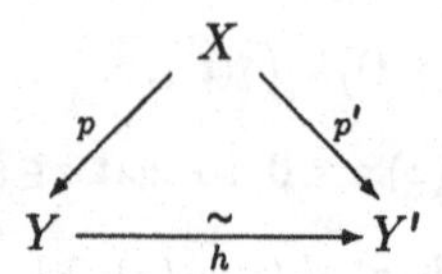

commutes; by (ii) this bijection is a homeomorphism.

Loosely speaking this means that up to isomorphism there is at most one definable quotient of X by E. If such a quotient (p, Y) exists, we are therefore justified in calling it *the* definable quotient of X by E, and write $Y = X/E$. We shall also use the phrase "$p: X \to Y$ is a definable quotient of X by E" instead of "(p, Y) is a definable quotient of X by E".

(3) A definable quotient $p: X \to Y$ is a quotient of X by E in the category of definable sets and continuous definable maps, in the sense that if $f: X \to Z$ is any continuous definable map into a definable set $Z \subseteq R^n$ such that $E \subseteq E_f$, then the unique map $g: Y \to Z$ such that $f = g \circ p$ is continuous and (obviously) definable.

(4) For a definable quotient of X by E to exist it is clearly necessary that E is closed in $X \times X$. But this is by no means sufficient, see Brumfiel [7, p. 71].

(2.4) EXAMPLE. Let $X \subseteq R^m$ be definable and $A \subseteq X$ a nonempty definable subset of X. We would like to "definably collapse" A to a point. In precise terms, consider the definable equivalence relation E_A on X whose equivalence classes are the singletons $\{x\}$ with $x \in X - A$, and the set A; we would like to find the definable quotient X/E_A. Concerning this we have the following result.

PROPOSITION. *If A is closed and bounded in R^m, then there is a definable proper quotient of X by E_A.*

PROOF. Consider the definable continuous map $p: X \to R^{m+1}$ given by $p(x) = (d_A(x) \cdot x, d_A(x))$, where $d_A(x) := \inf\{|x - a| : a \in A\}$ is the distance from x to A. Put $Y := p(X)$, and consider p as a map $X \to Y$. Assume A is closed and bounded. Claim: (p, Y) is a definably proper quotient of X by E_A. Clause (i) is clearly satisfied. To check the "proper" version of (ii), consider a definable curve $\gamma: (b, c) \to X$ such that $p(\gamma) \to p(x) \in Y$, $x \in X$. By Chapter 6, (4.5) it suffices to show that γ is completable in X. If $\lim_{t\to c} d_A(\gamma(t)) = 0$, then $\gamma \to a$ for some $a \in A$, since A is closed and bounded in R^m, and we are done. So we may assume that $\lim_{t\to c} d_A(\gamma(t)) = d_A(x) \neq 0$. In combination with $\lim_{t\to c} d_A(\gamma(t))\cdot\gamma(t) = d_A(x)\cdot x$, this gives $\lim_{t\to c} \gamma(t) = x$, and we are done again. □

This is a very special case of our main theorem (2.15), but before we can get to that we need a number of preliminaries, on completions, disjoint sums, and attaching spaces respectively.

(2.5) COMPLETIONS.

A **completion** of a definable set $S \subseteq R^m$ is a pair (h, S') consisting of a closed and bounded definable set $S' \subseteq R^n$ (for some n) and a definable map $h : S \to S'$ such that h is a homeomorphism from S onto $h(S)$ and $h(S)$ is dense in S'. We also express this more informally by: $h: S \to S'$ is a completion of S.

Clearly each definable set $S \subseteq R^m$ has a completion: just take a definable map $\mu: R^m \to R^m$ that maps R^m homeomorphically onto $(-1,1)^m$, put $S' :=$ closure of $\mu(S)$, and let $h: S \to S'$ be the restriction of μ. Then $h: S \to S'$ is a completion of S.

(2.6) Let $f: S \to T$ be a definable continuous map between definable sets $S \subseteq R^m$ and $T \subseteq R^n$. Then a **completion of** $f: S \to T$ is a triple consisting of a completion $i: S \to S'$ of S, a completion $j: T \to T'$, and a definable continuous map $f': S' \to T'$ such that

$$f' \circ i = j \circ f.$$

We also express this by saying that the commutative diagram

$$(*) \qquad \begin{array}{ccc} S & \xrightarrow{i} & S' \\ {\scriptstyle f}\downarrow & & \downarrow{\scriptstyle f'} \\ T & \xrightarrow[j]{} & T' \end{array}$$

is a **completion diagram** of $f: S \to T$.

There is always a completion of $f : S \to T$, but for our purpose we need a more precise result:

(2.7) LEMMA. *Let a completion $j : T \to T'$ be given. Then there is a completion diagram $(*)$ of $f : S \to T$ with the given map j as the bottom map in $(*)$. If in addition S is given as a definable closed subset of a definable set $X \subseteq R^m$, then there are a completion $h : X \to X'$ and a completion diagram $(*)$ of f, such that $S' =$ closure of $h(S)$ in X', and $i = h|S : S \to S'$.*

PROOF. Choose a completion $g : S \to S^C$ of S, and note that the definable map $g \times j : S \times T \to S^C \times T'$ is a homeomorphism onto its image $g(S) \times j(T)$. Let

S' be the closure of $(g \times j)(\Gamma(f))$ in $S^C \times T'$. Then $i : S \to S'$ given by $i(x) = (g \times j)(x, f(x)) = (g(x), jf(x))$ is a completion of S, and the restriction of the projection map $S^C \times T' \to T'$ to S' is a definable continuous map $f' : S' \to T'$ making the diagram $(*)$ commutative.

Suppose now that S is a definable closed subset of the definable set $X \subseteq R^m$. In the ambient affine space of T' we take a closed ball B that contains T'. By Chapter 8, (3.10) there is a definable continuous extension $\tilde{f}: X \to B$ of $j \circ f: S \to T'$. Apply the first part of the lemma to get a completion of $\tilde{f}: X \to B$ consisting of a completion $h: X \to X'$, the completion $1_B: B \to B$, and a definable continuous map $\tilde{f}': X' \to B$ such that $\tilde{f}' \circ h = \tilde{f}$. Put $S' :=$ closure of $h(S)$ in X', $i := h|S: S \to S'$, and note that $\tilde{f}'(S') \subseteq T'$. Hence $f' := \tilde{f}'|S'$ provides us with a completion diagram $(*)$ as required. □

(2.8) DISJOINT SUMS.

Let $S_1, \ldots, S_k$ be definable sets in $R^{m(1)}, \ldots, R^{m(k)}$, $k \geq 1$.

DEFINITION. A **disjoint sum** of $S_1, \ldots, S_k$ is a tuple $(h_1, \ldots, h_k, T)$ consisting of a definable set $T \subseteq R^n$ for some n, and definable maps $h_i: S_i \to T$ such that:

(i) h_i is a homeomorphism onto $h_i(S_i)$ and $h_i(S_i)$ is open in T, for $i = 1, \ldots, k$;
(ii) T is the disjoint union of the sets $h_1(S_1), \ldots, h_k(S_k)$.

Let $n = 1 + \max\{m(i) : 1 \leq i \leq k\}$ and define $h_i: S_i \to R^n$ by $h_i(x) := (x, i, \ldots, i)$. Then $(h_1, \ldots, h_k, \bigcup_i h_i(S_i))$ is clearly a disjoint sum of $S_1, \ldots, S_k$.

If $(h_1, \ldots, h_k, T)$ and $(h'_1, \ldots, h'_k, T')$ are both disjoint sums of $S_1, \ldots, S_k$, then there is a unique definable homeomorphism $\lambda : T \to T'$ such that $\lambda \circ h_i = h'_i$ for $i = 1, \ldots, k$. This means, loosely speaking, that there is up to isomorphism exactly one disjoint sum of $S_1, \ldots, S_k$. In the following we fix such a disjoint sum $(h_1, \ldots, h_k, T)$ of $S_1, \ldots, S_k$, and we write $S_1 \amalg \cdots \amalg S_k$ for T; there will be no harm in identifying each S_i with its image in $S_1 \amalg \cdots \amalg S_k$ via h_i.

(2.9) CONSTRUCTION OF $X \amalg_f Y$.

Let $X \subseteq R^m$ and $Y \subseteq R^n$ be definable sets, and let a definable continuous map $f: A \to Y$ from a definable set $A \subseteq X$ into Y be given. We would like to attach X to Y via f. To this end we consider on the disjoint sum $X \amalg Y$ the smallest equivalence relation $E(f)$ for which each $a \in A \subseteq X$ is equivalent to $f(a) \in Y$. One checks easily that

$$E(f) = \Delta(X) \cup \Delta(Y) \cup \{(a, f(a)) : a \in A\} \cup \{(f(a), a) : a \in A\}$$
$$\cup \{(a_1, a_2) \in A \times A : f(a_1) = f(a_2)\},$$

where $\Delta(X) = \{(x, x) : x \in X\}$ and $\Delta(Y) = \{(y, y) : y \in Y\}$ are the diagonals of X and Y. This shows that $E(f)$ is a *definable* equivalence relation; the question is whether the definable quotient of $X \amalg Y$ by $E(f)$ exists. If it does, we denote this quotient by $X \amalg_f Y$ rather than by $X \amalg Y/E(f)$, and we say that $X \amalg_f Y$ is the **space obtained by attaching X to Y via f.**

(2.10) UNIVERSAL PROPERTY OF $X \amalg_f Y$.

Suppose $X \amalg_f Y$ exists. Let $\xi: X \to X \amalg_f Y$ be the composition

$$X \to X \amalg Y \to X \amalg_f Y,$$

and similarly for $\eta: Y \to X \amalg_f Y$. Clearly ξ and η are definable and continuous and $\xi(a) = \eta(f(a))$ for all $a \in A$. Moreover, given any two definable continuous maps $\xi': X \to Z$ and $\eta': Y \to Z$ into a definable set Z with $\xi'(a) = \eta'(f(a))$ for all $a \in A$, there is exactly one map $\mu: X \amalg_f Y \to Z$ such that $\mu \circ \xi = \xi'$ and $\mu \circ \eta = \eta'$; this map μ is continuous and (obviously) definable.

(2.11) LEMMA. *Let A be closed and bounded in the ambient space R^m of X. Then $X \amalg_f Y$ exists as a definably proper quotient of $X \amalg Y$.*

PROOF. If $A = \emptyset$, then the identity map $X \amalg Y \to X \amalg Y$ is clearly a definably proper quotient of $X \amalg Y$ by $E(f)$. So let us assume $A \neq \emptyset$. Let R^M be the ambient space of $X \amalg Y$. We identify X and Y with their images in $X \amalg Y$; note that A is then closed and bounded in R^M. Let $d_A: R^M \to R$ be the distance function given by

$$d_A(z) := \min\{|z - a| : a \in A\}.$$

Let also $\tilde{f}: X \to R^M$ be a continuous definable extension of $f: A \to Y$. (Such $\tilde{f}$ exists by Chapter 8, (3.10).) Define the map $p: X \amalg Y \to R^{2M+1}$ by

$$p(x) = (\tilde{f}(x), d_A(x) \cdot x, d_A(x)) \text{ for } x \in X \text{ and } p(y) = (y, 0, 0) \text{ for } y \in Y,$$

so p is definable and continuous, and once checks easily that $E_p = E(f)$. Let $Z = p(X \amalg Y)$, a subset of R^{2M+1}. We will show that (p, Z) is the desired definably proper quotient of $X \amalg Y$ by $E(f)$. Let γ be a definable curve γ in $X \amalg Y$ such that $p(\gamma)$ is completable in Z.

By Chapter 6, (4.5) it suffices to show that γ is completable in $X \amalg Y$. We may as well assume (after restricting the domain of γ suitably) that either γ lies entirely in X, or γ lies entirely in Y. In the first case we may further assume γ lies entirely in $X - A$ and is bounded away from A since otherwise γ will be completable in X with $\gamma \to a \in A$; with this further assumption on γ and using the assumption that $p(\gamma)$ is completable in Z it follows just as in the proof of (2.4) that γ is completable in X. The case that γ lies entirely in Y is even simpler and left to the reader. □

We now use completions to extend this lemma as follows:

(2.12) PROPOSITION. *Suppose A is closed in X and $f: A \to Y$ is definably proper. Then $X \amalg_f Y$ exists as a definably proper quotient of $X \amalg Y$.*

PROOF. Identifying X and Y with their images in suitable completions and using lemma (2.7) we may assume that X and Y are bounded in their ambient space and f extends to a definable continuous map $\mathrm{cl}(f) : \mathrm{cl}(A) \to \mathrm{cl}(Y)$. Since f is definably proper we have $\mathrm{cl}(f)^{-1}(Y) = A$. (To see this, let $x \in \mathrm{cl}(A)$, $\mathrm{cl}(f)(x) \in Y$. To derive $x \in A$ we take a definable path $\gamma : [0,1] \to \mathrm{cl}(A)$ with $\gamma(1) = x$ and $\gamma([0,1)) \subseteq A$. Then $\mathrm{cl}(f)(\gamma([0,1])) \subseteq Y$ is closed and bounded in the ambient affine space of Y, so $f^{-1}(\mathrm{cl}(f)(\gamma([0,1])))$ is a subset of A and is closed and bounded in the ambient affine space of X. Since this set contains $\gamma([0,1])$ it must also contain $\gamma(1) = x$, hence $x \in A$.) By (2.11) we have a definably proper quotient $\mathrm{cl}(p) : \mathrm{cl}(X) \amalg \mathrm{cl}(Y) \to X \amalg_{\mathrm{cl}(f)} Y$. Consider $X \amalg Y$ as a subset of $\mathrm{cl}(X) \amalg \mathrm{cl}(Y)$ in the obvious way and let $Z := \mathrm{cl}(p)(X \amalg Y)$. It is easily seen that $\mathrm{cl}(p)^{-1}(Z) = X \amalg Y$ and that $E(\mathrm{cl}(f)) \cap (X \amalg Y)^2 = E(f)$. Hence $p := \mathrm{cl}(p)|X \amalg Y : X \amalg Y \to Z$ is a definably proper quotient of $X \amalg Y$ by $E(f)$. □

(2.13) Let E be a definable equivalence relation on a definable set X. Let $pr_1 : E \to X$ and $pr_2 : E \to X$ be the restrictions of the two projection maps $X \times X \to X$.

We call E **definably proper over** X if pr_1 (equivalently, pr_2) is a definably proper map. One checks that if X/E exists as a definably proper quotient of X by E, then E is definably proper over X. The main theorem (2.15) below asserts the converse. We need one more lemma before embarking on its proof.

(2.14) LEMMA. *Let $E \subseteq X \times X$ be a definable equivalence relation on a definable set $X \subseteq R^m$ and suppose E is definably proper over X. Then, given a definable curve $\gamma = (\alpha, \beta)$ in E with $\alpha = pr_1(\gamma)$, $\beta = pr_2(\gamma)$, if either α or β is completable in X, both are completable in X, and in that case, if $\alpha \to p$, $\beta \to q$, then $\gamma \to (p, q) \in E$. (Hence E is closed in $X \times X$.)*

PROOF. Suppose α is completable in X. Since $pr_1 : E \to X$ is definably proper and γ lifts α, it follows that γ is completable in E. Hence $\beta = pr_2(\gamma)$ is completable in X. □

(2.15) THEOREM. *Suppose the definable equivalence relation E on the definable set X is definably proper over X. Then X/E exists as a definably proper quotient of X.*

This contains proposition (2.12) on $X \amalg_f Y$ as a special case, but (2.12) is needed in the proof of (2.15). Here are two further important special cases:

(2.16) COROLLARY. *If $X \subseteq R^m$ is closed and bounded and $E \subseteq X \times X$ is a closed definable equivalence relation, then X/E exists as a definably proper quotient of X.*

(2.17) Application to orbit spaces. Let G be a definable topological group, that is, G is a definable subset of some R^m equipped with a definable continuous group operation $G \times G \to G$. (We leave it as an exercise to show that then the group inversion $x \mapsto x^{-1} : G \to G$ is also continuous, so that G is indeed a topological group.) Let in addition a continuous definable group action $G \times X \to X$ on a definable set X be given. Then the orbits Gx are exactly the equivalence classes of the definable equivalence relation $\sim_G$ on X given by $x \sim_G y \Leftrightarrow gx = y$ for some $g \in G$.

(2.18) Corollary. *If G is closed and bounded (for example, finite), then the orbit space $X/G := X/\sim_G$ exists as a definably proper quotient of X.*

(2.19) Proof of theorem (2.15). By induction on $\dim(X)$. If $\dim(X) \leq 0$, then X is finite; in that case the theorem holds trivially. Assume $\dim(X) = d > 0$. By Chapter 6, (1.2), there is a definable set $S \subseteq X$ that has exactly one point in common with each equivalence class of E. Let $\sigma : X \to S$ be the definable map that assigns to each $x \in X$ the unique point in S to which it is equivalent; σ may not be continuous, let alone definably proper. Nevertheless we can use σ and S to construct X/E. Let $c(S)$ be the closure of S in X. Let $B = \sigma\big(c(S) - S\big)$ be the set of points of S that are equivalent to points of $c(S) - S$, so $\dim(B) < d$. Take a triangulation of X that is compatible with S, $c(S)$ and B, let $\mathcal{P}$ be the partition of X corresponding to this triangulation, and let S_d be the union of the d-dimensional sets of $\mathcal{P}$ that are contained in S. One checks easily that S_d is open in X. Put $S' := c(S) - S_d$. Since $B \cap S_d = \emptyset$, no point of S_d is equivalent to a point of S'. Moreover S' is closed in X. Hence $E' := E \cap (S' \times S')$ is definably proper over S'. Since $\dim(S') < d$ there is by the inductive hypothesis a definably proper map $f' : S' \to Y'$ onto a definable set Y' with $E' = E_{f'}$. Note that f' maps $S - S_d$ bijectively onto Y'. Let $c(S_d)$ be the closure of S_d in X, and put $A := c(S_d) \cap S'$. We construct $Y = X/E$ by attaching $c(S_d)$ to Y' via $f'' := f'|A : A \to Y'$. Note that A is closed in $c(S_d)$, and that $f'' : A \to Y'$ is definably proper, since f' is. Hence we can apply proposition (2.12) to obtain $Y := c(S_d) \amalg_{f''} Y'$ as a definably proper quotient of $c(S_d) \amalg Y'$ via the map $p : c(S_d) \amalg Y' \to Y$.

Note that the composed map $S' \to Y' \to Y$ agrees with the map $c(S_d) \to Y$ on the intersection A of their domains, hence these two maps determine a (definable, continuous) map $g : c(S) = c(S_d) \cup S' \to Y$. Consider the following commuting diagram:

$$\begin{array}{ccc} c(S_d) \amalg S' & \xrightarrow{\ j\ } & c(S_d) \amalg Y' \\ {\scriptstyle h}\downarrow & & \downarrow{\scriptstyle p} \\ c(S) & \xrightarrow{\ g\ } & Y = c(S_d) \amalg_{f''} Y' \end{array}$$

where the (continuous, definable) map j is the identity on $c(S_d)$ and equal to f' on S', and the (continuous, definable) map h is induced by the inclusion maps $c(S_d) \to c(S)$ and $S' \to c(S)$. Note that all four maps are surjective, that p is definably proper, and that j is definably proper since f' is. An easy diagram chase then shows that g is definably proper. Since f' maps $S - S_d$ bijectively onto Y', the map $g|S: S \to Y$ is a bijection. Now put $f := g \circ \sigma : X \to S \to Y$. Clearly f is definable, surjective, and $E = E_f$. As mentioned already, σ may not be continuous, but we claim

$$(*) \qquad f: X \to Y \text{ is continuous and definably proper.}$$

(Clearly $(*)$ implies that $f: X \to Y$ is a definably proper quotient of X by E, so we have reduced to proving $(*)$.) To check continuity of f, take a definable curve α in X with $\alpha \to p \in X$. Then $(\alpha, \sigma(\alpha))$ is a definable curve in E and α is completable in X, hence $\sigma(\alpha)$ is completable in X by lemma (2.14), say $\sigma(\alpha) \to q \in c(S)$. Then $(p, q) \in E$, and $f(\alpha) = g(\sigma(\alpha)) \to g(q)$. But $g(q) = f(q)$: if $q \in S$, this follows from $\sigma(q) = q$; if $q \in c(S) - S$, then $\sigma(q) \in S - S_d$, and both q and $\sigma(q)$ are in S', so that $f'(q) = f'(\sigma(q))$, hence $g(q) = g(\sigma(q)) = f(q)$. Also $f(q) = f(p)$. Hence $f(\alpha) \to f(p)$. This proves continuity of f. To show f is definably proper, take a definable curve α in X such that $f(\alpha) = g(\sigma(\alpha))$ is completable in Y. Since g is definably proper, $\sigma(\alpha)$ is completable in $c(S)$. Again by lemma (2.14) it follows that α is completable in X. □

Notes and comments

The notion of definable space (even for arbitrary o-minimal structures) is implicit in Pillay [48]. It is used there to study the properties of groups and fields that are definable in o-minimal structures. See also Khovanskii [33] for similar gluing constructions in the Pfaffian setting.

For the proof of theorem (1.8) I largely followed Robson's treatment in [50] of the semialgebraic case. The material on completions in Section 2 is adapted from Delfs and Knebusch [12], and the remainder of Section 2 follows Brumfiel [8].

For o-minimal expansions of "the real field with restricted analytic functions" there is also a quite different way to go beyond the affine setting, by introducing a (locally defined) notion of "nice" subset of an arbitrary real analytic manifolds, generalizing "subanalytic" sets; see [23].

HINTS AND SOLUTIONS

Chapter 2, (3.7)

3. Write R^m as a disjoint union of semialgebraic sets $A_0, A_1, \ldots, A_k$ with $A_1, \ldots, A_k$ connected such that if $x \in A_0$, then $Q(x,T) = 0$, while if $1 \leq i \leq k$ and $x \in A_i$, then the real roots of $Q(x,T)$ are given by continuous semialgebraic functions $\zeta_{i1}(x) < \cdots < \zeta_{ie(i)}(x)$ of x. Note that then g must coincide on each A_i $(i > 0)$ with one of the functions ζ_{ij}. Next use the fact that each point in A_0 is in the closure of $A_1 \cup \cdots \cup A_k$.

Chapter 3, (2.19)

3. Find a partition of C into cells $C_1, \ldots, C_k$ and for each $i = 1, \ldots, k$ find continuous definable functions $\gamma_{i0}, \ldots, \gamma_{ir(i)} : C_i \to R$ with

$$\alpha | C_i = \gamma_{i0} < \gamma_{i1} < \cdots < \gamma_{ir(i)} = \beta | C_i,$$

such that for each cell $\Gamma(\gamma_{ij})$ $(1 \leq j < r(i))$ and each cell $(\gamma_{ij}, \gamma_{ij+1})$ $(0 \leq j < r(i))$:

(i) the restriction of f to $\Gamma(\gamma_{ij})$ is continuous,

(ii) the restriction of f to $(\gamma_{ij}, \gamma_{ij+1})$ is continuous, and either strictly increasing in the $(m+1)^{\text{th}}$ variable, or independent of the $(m+1)^{\text{th}}$ variable, or strictly decreasing in the $(m+1)^{\text{th}}$ variable.

Then one shows along the lines of the proof of lemma (2.16) of Chapter 3 that each restriction $f | (\gamma_{i0}, \gamma_{ir(i)})$ is continuous.

Chapter 3, (3.8)

Given an L-formula $\phi(x_1, \ldots, x_m, y)$, there are natural numbers M and N such that for all $r \in R^m$ the set $\phi(r, \mathcal{R}) \subseteq R$ is a union of at most M intervals and at most N points. This property is inherited by $\mathcal{R}'$.

Chapter 4, (1.17)

3. Use the canonical homeomorphism of A with an open cell to show that a nonempty definable open subset of A has dimension d.

4. Partition A into finitely many cells, and note that $\{a \in \mathrm{cl}(A) : \dim_a(A) < d\}$ is contained in the union of the closures of the cells of dimension $< d$ in this partition.

Chapter 6, (1.15)

2 and 3. With $T \subseteq X$ as in exercise 1, let $f: X \to T$ be the map assigning to each x its representative in T. Now apply (1.6)(ii) and (2.11) from Chapter 4.

4. If for a certain $\epsilon > 0$ there is no such δ, find definable continuous maps $\gamma_1, \gamma_2 : (0, a] \to X$ for some $a > 0$ such that $|f(\gamma_1(t)) - f(\gamma_2(t))| \geq \epsilon$ for all $t \in (0, a]$, and $\lim_{t \to 0} |\gamma_1(t) - \gamma_2(t)| = 0$.

5. Consider points in X where the function $x \mapsto |f(x) - x| : X \to R$ takes its minimum value.

Chapter 6, (4.8)

2. Suppose f is definably proper with respect to $(R, <, \mathcal{S})$, and let α be a curve in X definable in $(R, <, \mathcal{S}')$, such that $f(\alpha) \to y \in Y$. Since Y is locally closed, y has a neighborhood N in Y that is definable in $(R, <, \mathcal{S})$ and closed and bounded in R^n. To show α is completable in X we may as well assume that $f(\alpha)$ lies entirely in N. Then α lies in $f^{-1}(N)$ and $f^{-1}(N)$ is closed and bounded in R^m, so α is completable in $f^{-1}(N)$, and hence in X.

3. Suppose f is definably proper. Let $S \subseteq X$ be closed in X. To show $f(S)$ is closed in Y, write $S = \bigcap_i S_i$ where $(S_i)_{i \in I}$ is a family of definable closed subsets of X such that for any two indices i_1 and i_2 there is an index i with $S_i \subseteq S_{i_1} \cap S_{i_2}$. It suffices to show that $f(S) = \bigcap_i f(S_i)$. Let $q \in \bigcap_i f(S_i)$, so that $f^{-1}(q) \cap S_i$ is a closed nonempty subset of $f^{-1}(q)$ for all i. Since $f^{-1}(q)$ is compact, it follows that the sets $f^{-1}(q) \cap S_i$ have a common point p, so $p \in S$ and $f(p) = q$.

Chapter 7, (4.3)

1. Let $\phi: R^{m+1} \to \{x \in R^{m+1} : \|x\| < 1\}$ be the definable homeomorphism given by $\phi(x) = x/\sqrt{1 + \|x\|^2}$. Use that $u \in S^m$ is an asymptotic direction for A if and only if $u \in \mathrm{cl}(\phi(A))$. Then apply theorem (1.8) from Chapter 4 to $\phi(A)$.

3. Use the fact that a basic semilinear set in R^m is an open subset of an affine subspace of R^m, and that R^m cannot be covered by finitely many linear subspaces of dimension $< m$.

Chapter 8, (2.14)

2. (Bounded semilinear sets are polyhedrons.) Along the same lines as the proof of the triangulation theorem. Assume the desired result holds for a certain value of m, and let $S, S_1, \ldots, S_k$ be bounded semilinear sets in R^m, $S_i \subseteq S$. Replacing S by its closure (and adding the old S to the list of distinguished subsets) we may as well assume that S is closed. Let $T := \mathrm{bd}(S) \cup \mathrm{bd}(S_1) \cup \cdots \cup \mathrm{bd}(S_k)$, so T is closed and bounded of dimension $< m + 1$. After applying a linear automorphism of R^{m+1} we may assume by exercise 3 of Chapter 7, (4.3) that e_{m+1} is a good direction vector for T.

By the inductive hypothesis, and by Chapter 1, (7.4), there are a closed complex K in R^m and a finite collection F of affine functions on R^m, such that $|K| = \pi(T)$, the restrictions to each $\sigma \in K$ of the functions in F can be arranged in increasing order as $f_{\sigma,1} < \cdots < f_{\sigma,j(\sigma)}$, and $T = \bigcup\{\Gamma(f_{\sigma,j}) : \sigma \in K,\ 1 \leq j \leq j(\sigma)\}$. We also take care that K is compatible with all sets $\pi\bigl(S \cap \Gamma(f)\bigr)$ and $\pi\bigl(S_i \cap \Gamma(f)\bigr)$ $(f \in F,\ 1 \leq i \leq k)$. We now indicate how to use the arguments in the proof of lemma (2.8) to get a complex L in R^{m+1} such that $K = \{\pi(\sigma) : \sigma \in L\}$, $|L| = S$ and each S_i is a union of simplexes of L.

Indeed, let $A \subseteq R^m$ be any closed and bounded semilinear set, F any finite nonempty collection of affine functions on R^m, and K any closed complex in R^m with $A = |K|$, such that for each $\sigma \in K$ the restrictions of the functions in F to σ can be arranged in increasing order as $f_{\sigma,1} < \cdots < f_{\sigma,j(\sigma)}$. (It may happen that different functions in F have the same restriction to σ.)

Given two successive functions $f = f_{\sigma,i}$ and $g = f_{\sigma,i+1}$ we note that for at least one vertex p of σ we must have $\mathrm{cl}(f)(p) < \mathrm{cl}(g)(p)$, where $\mathrm{cl}(f), \mathrm{cl}(g)$ denote the continuous extensions of f and g to $\mathrm{cl}(\sigma)$. Fixing a linear order on $\mathrm{Vert}(K)$ we obtain in this way for each successive pair f, g as above a closed complex $L(f,g)$ in R^{m+1} such that $|L(f,g)|$ is the convex hull of

$$\{\mathrm{cl}(f)(p) : p \text{ a vertex of } \sigma\} \cup \{\mathrm{cl}(g)(p) : p \text{ a vertex of } \sigma\},$$

which equals $[\mathrm{cl}(f), \mathrm{cl}(g)]$, and also for each $f = f_{\sigma,i}$ a closed complex $L(f)$ in R^{m+1} with $|L(f)| = \Gamma\bigl(\mathrm{cl}(f)\bigr)$.

All this is just as in the proof of lemma (2.8). Finally, the union of the $L(f,g)$'s and $L(f)$'s is a closed complex in R^m, and this complex is compatible with the sets in $K^F :=$ set of the (f,g)'s and $\Gamma(f)$'s.

3. Let $\mathcal{R}$ be an L-structure, where L extends the language of ordered rings. We may as well assume that S is defined by an L-formula $\phi(x,y)$, $x = (x_1,\ldots,x_m)$, $y = (y_1,\ldots,y_n)$, since the constants from R that may be involved in an $L(R)$-formula defining S can be replaced by extra parametric variables (increasing m).

Let $\mathcal{R}'$ be any L-structure elementarily equivalent to $\mathcal{R}$. Then $\mathcal{R}'$ is also o-minimal, by Chapter 3, (3.8), and hence satisfies triangulation. Therefore, given any $a \in R'^m$ there is a definable homeomorphism h_a from $\phi(a, R'^n)$ onto a union of faces of the simplex $(e_1,\ldots,e_N)$ in R'^N, for some $N = N(a) \in \mathbf{N}$. Because $\mathrm{Th}(\mathcal{R})$ has definable Skolem functions, we may further assume the graph of h_a is defined by a formula $\psi(a,y,z)$ where $\psi(x,y,z)$ is an L-formula depending on a and with $z = (z_1,\ldots,z_N)$. Since $\mathcal{R}' \models \mathrm{Th}(\mathcal{R})$ was arbitrary it follows by model-theoretic compactness that only finitely many different such formulas ψ and numbers $N(a) \in \mathbf{N}$ are needed when $\mathcal{R}'$ varies over all models of $\mathrm{Th}(\mathcal{R})$ and a over R'^m. We can easily construct from these finitely many ψ's a single definable map f as required, with N the maximum of the finitely many numbers $N(a)$.

REFERENCES

1. E. Artin and O. Schreier, *Algebraische Konstruktion reeller Körper*, Hamb. Abh. **5** (1926), 85–99.

2. R. Benedetti and J.-J. Risler, *Real algebraic and semi-algebraic sets*, Hermann, Paris, 1990.

3. E. Bierstone and P. Milman, *Semianalytic and subanalytic sets*, IHES Publ. Math. **67** (1988), 5–42.

4. J. Bochnak, M. Coste and M.-F. Roy, *Géométrie algébrique réelle*, Ergebnisse der Math. und ihrer Grenzgebiete, 3. Folge, Band 12, Springer, Berlin-Heidelberg, 1987.

5. J. Bochnak and G. Efroymson, *Real algebraic geometry and the Hilbert* 17th *problem*, Math. Ann. **251** (1980), 213–241.

6. N. Bourbaki, *Elements of Mathematics, General Topology, Part 1*, Hermann, Paris, 1966.

7. G. Brumfiel, *Partially ordered rings and semi-algebraic geometry*, Cambridge University Press, Cambridge, 1979.

8. ———, *Quotient spaces for semialgebraic equivalence relations*, Math. Z. **195** (1987), 69–78.

9. C. Chang and H. Keisler, *Model Theory*, third edition, North-Holland, Amsterdam, 1990.

10. M. Coste and M. Shiota, *Nash triviality in families of Nash manifolds*, Inv. Math. **108** (1992), 349–368.

11. H. Delfs and M. Knebusch, *Homology of algebraic varieties over real closed fields*, J. reine angew. Math. **335** (1982), 122–163.

12. ———, *Locally Semialgebraic Spaces*, Springer Lecture Notes 1173, Springer, Berlin, 1985.

13. ———, *Separation, retractions and homotopy extension in semialgebraic spaces*, Pac. J. Math. **114** (1984), 47–71.

14. C. Delzell, *A finiteness theorem for open semi-algebraic sets, with applications to Hilbert's 17th problem*, in "Ordered Fields and Real Algebraic Geometry", D.W. Dubois and T. Recio, eds., Contemp. Math. 8, AMS, 1982, pp. 79–97.

15. J. Denef and L. van den Dries, *p-Adic and real subanalytic sets*, Ann. Math. **128** (1988), 79–138.

16. J. Dieudonné, *A History of Algebraic and Differential Topology 1900–1960*, Birkhäuser, Boston, Mass., 1989.

17. L. van den Dries, *Some applications of a model theoretic fact to (semi-) algebraic geometry*, Indag. Math. **44** (1982), 397–401.

18. ———, *Algebraic theories with definable Skolem functions*, J. Symb. Logic **49** (1984), 625–629.

19. ———, *Remarks on Tarski's problem concerning* $(\mathbb{R}, +, \cdot, \exp)$, in "Logic Colloquium '82", G. Lolli, G. Longo and A. Marcja, eds., North-Holland, 1984, pp. 97–121.

20. ———, *A generalization of the Tarski-Seidenberg theorem, and some nondefinability results*, Bull. AMS **15** (1986), 189–193.

21. ———, *O-minimal structures*, in "Logic: from Foundations to Applications (Conference Proceedings)", W. Hodges et al., eds., Oxford University Press, 1996, pp. 137–185.

22. L. van den Dries, A. Macintyre and D. Marker, *The elementary theory of restricted analytic fields with exponentiation*, Ann. Math. **140** (1994), 183–205.

23. L. van den Dries and C. Miller, *Geometric categories and o-minimal structures*, Duke Math. J. **84** (1996), 497-540.

24. R. Dudley, *A course on empirical processes*, in "École d'Été de Probabilités de Saint-Flour 1982"; Lecture Notes in Math., vol. 1097, Springer, Berlin, 1984, pp. 1–142.

25. L. Fuchs, *Partially ordered algebraic structures*, Pergamon Press, Oxford, 1963.

26. A. Gabrielov, *Projections of semi-analytic sets*, Funct. Anal. Appl. **2** (1968), 282–291.

27. B. Giesecke, *Simpliziale Zerlegung abzählbarer analytischer Räume*, Math. Z. **83** (1964), 177–213.

28. A. Grothendieck, *Esquisse d'un Programme*, Research Proposal (unpublished) (1984).

29. R. Hardt, *Semi-algebraic local triviality in semi-algebraic mappings*, Amer. J. Math. **102** (1980), 291–302.

30. H. Hironaka, *Introduction to real-analytic sets and real-analytic maps*, Lecture Notes, Istituto Matematico "L. Tonelli", Pisa, 1973.

31. ———, *Triangulation of algebraic sets*, in "Algebraic Geometry"; Proc. Symp. Pure Math., vol. 29, AMS, Providence, RI, 1975, pp. 165–185.

32. A. Khovanskii, *On a class of systems of transcendental equations*, Sov. Math. Dokl. **22** (1980), 762–765.

33. ———, *Fewnomials*, Translations of Mathematical Monographs, vol. 88, AMS, Providence, RI, 1991.

34. M. Knebusch, *Semialgebraic topology in the last ten years*, in "Real Algebraic Geometry Proceedings, Rennes 1991", M. Coste, L. Mahé, M.-F. Roy, eds., Lecture Notes in Math. 1524, Springer, 1992, pp. 1–36.

35. J. Knight, A. Pillay and C. Steinhorn, *Definable sets in ordered structures. II*, Trans. AMS **295** (1986), 593–605.

36. B. Koopman and A. Brown, *On the covering of analytic loci by complexes*, Trans. AMS **34** (1932), 231–251.

37. S. Lang, *Algebra*, second edition, Addison-Wesley, Reading, Mass., 1984.

38. C. Laskowski, *Vapnik-Chervonenkis classes of definable sets*, J. London Math. Soc. **245** (1992), 377–384.

39. S. Łojasiewicz, *Triangulation of semi-analytic sets*, Ann. Scuola Norm. Sup. Pisa **18** (1964), 449–474.

40. ———, *Ensembles semi-analytiques*, Lecture Notes, IHES, Bures-sur-Yvette, France, 1965.

41. A. Macintyre and E. Sontag, *Finiteness results for sigmoidal "neural" networks*, in "Proc. 25th Annual Symp. Theory Computing", San Diego, May 1993, pp. 325–334.

42. R. MacPherson, *Intersection homology and perverse sheaves*, Colloquium Lectures (1991), AMS.

43. C. Miller, *Exponentiation is hard to avoid*, Proc. AMS. **122** (1994), 257–259.

44. ———, *Expansions of the real field with power functions*, Ann. Pure Appl. Logic **68** (1994), 79–94.

45. Y. Peterzil, *A structure theorem for semibounded sets in the reals*, J. Symb. Logic **57** (1992), 779–794.

46. Y. Peterzil and S. Starchenko, *A trichotomy theorem for o-minimal structures*, Proc. London Math. Soc. (to appear).

47. Y. Peterzil and C. Steinhorn, *Definable compactness and definable subgroups of o-minimal groups*, J. London Math. Soc. (to appear).

48. A. Pillay, *On groups and fields definable in o-minimal structures*, J. Pure Appl. Algebra **53** (1988), 239–255.

49. A. Pillay and C. Steinhorn, *Definable sets in ordered structures I*, Trans. AMS **295** (1986), 565–592.

50. R. Robson, *Embedding semialgebraic spaces*, Math. Z. **184** (1983), 365–370.

51. S. Schanuel, *Negative sets of Euler characteristic and dimension*, in "Category Theory, Como 1990", A. Carbonari, M. Pedicchio, G. Rosonlini, eds., Lecture Notes in Math. 1488, Springer, 1991, pp. 379–385.

52. P. Scowcroft and L. van den Dries, *On the structure of semialgebraic sets over p-adic fields*, J. Symb. Logic **53** (1988), 1138–1164.

53. S. Shelah, *Stability, the f.c.p. and superstability*, Ann. Math. Logic **3** (1971), 271–362.

54. M. Shiota, *Geometry of Subanalytic and Semialgebraic Sets: Abstract*, in "Real analytic and algebraic geometry", F. Broglia et al., ed., W. de Gruyter, Berlin, 1995, pp. 251–275.

55. E. Sontag, *Remarks on piecewise-linear algebra*, Pac. J. Math. **98** (1982), 183–201.

56. ———, *Critical points for least-squares problems involving certain analytic functions, with applications to sigmoidal nets*, Adv. Comp. Math. **5** (1996), 245–268.

57. P. Speissegger, *Fiberwise properties of definable sets and functions in o-minimal structures*, Manuscripta Math. **86** (1995), 283–291.

58. G. Stengle and J. Yukich, *Some new Vapnik-Chervonenkis classes*, Ann. Statist. **17** (1989), 1441–1446.

59. A. Strzebonski, *Euler characteristic in semialgebraic and other o-minimal groups*, J. Pure Appl. Algebra **96** (1994), 173–201.

60. H. Sussmann, *Real analytic desingularization and subanalytic sets: An elementary approach*, Trans. AMS **317** (1990), 417–461.

61. A. Tarski, *A Decision method for Elementary Algebra and Geometry*, second edition revised, Rand Corporation, Berkeley and Los Angeles, 1951.

62. V. Vapnik and A. Chervonenkis, *On the uniform convergence of relative frequencies of events to their probabilities*, Th. Prob. Appl. **16** (1971), 264–280.

63. H. Whitney, *Elementary structure of real algebraic varieties*, Ann. Math. **66** (1957), 545–556.

64. A. Wilkie, *Model completeness results for expansions of the real field by restricted Pfaffian functions and the exponential function*, J. AMS **9** (1996), 1051–1094.

65. A. Woerheide, *O-minimal homology*, Ph.D. thesis (1996), University of Illinois at Urbana-Champaign.

INDEX

Printed in the United States
By Bookmasters